GALILEO REVISITED

DATE DUE

			PRINTED IN U.S.A.

PASCHAL SCOTTI

GALILEO
REVISITED

The Galileo Affair in Context

IGNATIUS PRESS   SAN FRANCISCO

Cover art:
Portrait of Galileo Galilei, 1636
Justus Sustermans (1597–1681)
National Maritime Museum, Greenwich, London
Wikimedia Commons Image

Galileo before the Holy Office
Joseph-Nicolas Robert-Fleury (1797–1890)
Wikimedia Commons Image

Cover design by Enrique Javier Aguilar Pinto

ISBN 978-1-62164-132-2
Library of Congress Control Number 2017932706
Printed in the United States of America ∞

To my valued friends and colleagues
Dom Edmund Adams and J. Clifford Hobbins

CONTENTS

INTRODUCTION

Galileo loved the fray. Not for him the laborious hours of observation of a Tycho; not for him the endless calculations and curve-fittings of a Kepler. He was a man with a vision of the way the universe had to be, and a talent for communicating that vision to others. None knew better than he that, whereas theorems have to be proved, people have to be persuaded.

—Ernan McMullin, *Galileo, Man of Science**

Galileo is one those iconic figures in history for whom there is endless fascination. Besides an abundant scholarly literature, the "Galileo industry", as one author put it, there also has been great general interest in him, as we see in a stream of biographies that still come out and in the international bestseller of more than a dozen years ago, Dava Sobel's *Galileo's Daughter*,[1] which sparked the 2002 PBS/Nova television program *Galileo's Battle for the Heavens*. Nor have many scientists inspired serious works of art, as we see with Philip Glass' 2002 opera *Galileo Galilei* and Bertold Brecht's *Galileo*, revived again in early 2012, off-Broadway, with F. Murray Abraham in the title role. Galileo's significance lies far beyond his importance in the history of science. Invariably, it is his brush with the Inquisition, his trial and condemnation by the Catholic Church in 1633, that seizes the imagination. Over and over again, it is the "Galileo Affair" that people remember. So much so, that when *National Geographic* put out a children's book on Galileo in 2005, it could not but entitle the book *Galileo: The Genius Who Faced the Inquisition*.

All of us have grown up with the idea of the warfare between science and religion. While the idea began with the Enlightenment, it reached its high-water mark in the Victorian period. Thomas Huxley

---

*Ernan McMullin, ed., *Galileo, Man of Science* (New York: Basic Books, 1967), 3.
[1]Dava Sobel, *Galileo's Daughter: A Historical Memoir of Science, Faith, and Love* (New York: Walker, 1999).

(1825–1895), the "Pope of Science", pushed the military metaphor in his attempt to professionalize science, moving it from a part of Christian apologetics, the preserve of Anglican gentlemen and clerics, or rather, gentlemen clerics, into a hard-edged secular discipline financed by the state, and ordered to public usefulness.[2] Only by discrediting the religious culture of the traditional, Anglican-landed Establishment could *his* science come into its own. Who can forget his pugnacious line, "Extinguished theologians lie about the cradle of every science as the strangled snakes beside that of Hercules."[3] This polarization, or at least the attempt to create polarization, was part of the means by which he achieved it; and it was as much about class, power, and prestige as about the pursuit of truth.[4] This was equally true of the United States, which, while not having an official Protestant Establishment, was a society profoundly influenced by the Protestant churches and where the clergy were among its cultural and political leaders. John William Draper's *History of the Conflict Between Religion and Science* (1874), and particularly Andrew Dickson White's *A History of the Warfare of Science with Theology in Christendom* (1896), were the great literary landmarks of this warfare genre, and in the latter Galileo has a prominent and honored place as a "martyr of science", with some forty pages devoted to the affair.[5]

While still common among journalists and much of the public, among scholars the "warfare" imagery is not as dominant as it once was: one scholar has written that while the reports of its death "may yet be premature, it seems safe to say that it has had a near-death experience, at least among professional historians of science".[6] Oddly enough, there has been something of a revival of the conflict model in reference to the Galileo case beginning in the 1990s, particularly among some Catholic scholars, but we will hear more of that later.

[2] Sheridan Gilley and Ann Loades, "Thomas Henry Huxley: The War between Science and Religion", *Journal of Religion* 61 (1981): 285–308.

[3] Adrian Desmond, *Huxley: From Devil's Disciple to Evolution's High Priest* (Reading, Mass.: Addison-Wesley, 1997), 292.

[4] Frank M. Turner, "The Victorian Conflict between Science and Religion: A Professional Dimension", *Isis* 69 (1978): 356–76.

[5] David C. Lindberg and Ronald L. Numbers, "Beyond War and Peace: A Reappraisal of the Encounter between Christianity and Science", *Church History* 55 (1986): 338–54.

[6] Edward Davis, "Christianity and Early Modern Science: The Foster Thesis Reconsidered", in *Evangelicals and Science in Historical Perspective*, ed. David N. Livingstone, D. G. Hart, and Mark A. Noll (Oxford: Oxford University Press, 1999), 75.

Recent scholarship has been much more positive about the Church's role in science. The respected historian of science Edward Grant rather sees Christianity as supportive of science and the Christian Middle Ages as laying the foundations for the Scientific Revolution.[7] Despite the clear religious orientation of the Middle Ages, science was given enormous institutional support in that uniquely medieval creation, the universities, where the arts curriculum was basically a scientific one and whose main job was the training of clerics. This respect for science was equally true of astronomy as of any other science. As J.L. Heilbron put it in his study of churches as solar observatories: "The Roman Catholic Church gave more financial and social support to the study of astronomy for over six centuries, from the recovery of ancient learning during the late Middle Ages into the Enlightenment, than any other, and, probably, all other, institutions."[8]

Richard Blackwell wrote in his 2006 book on Galileo's trial that Galileo's encounter with the Catholic Church was a major turning point in the history of Western culture and the defining event for the relationship between science and religion ever since.[9] Richard Westfall, in an earlier essay, has called it one of the climactic events of the Scientific Revolution and indeed of all European history.[10] Those are grand claims from two highly respected scholars, and they are certainly worthy of discussion. It is my hope that this book will shed some light on this complex reality. The Galileo case is inescapable, with its myths and countermyths. But as a pair of Gifford Lecturers have remarked, reflecting on the Galileo case, the intervention of the historian not only makes the story fuller; it offers the possibility of some surprising twists and turns, breaking the mold of our expectations and showing its contemporary relevance.[11]

In history context is everything. The Galileo Affair cannot be understood without understanding its context. While it has been

[7] Edward Grant, *The Foundations of Modern Science in the Middle Ages: Their Religious, Institutional and Intellectual Contexts* (Cambridge: Cambridge University Press, 1996), 168–206.

[8] J.L. Heilbron, *The Sun in the Church* (Cambridge, Mass.: Harvard University Press, 1999), 3.

[9] Richard Blackwell, *Behind the Scenes at Galileo's Trial* (Notre Dame, Ind.: Notre Dame University Press, 2006), 1.

[10] Richard Westfall, *Essays on the Trial of Galileo* (Notre Dame, Ind.: Vatican Observatory Publications, 1989), 1.

[11] John Brooke and Geoffrey Cantor, *Reconstructing Nature: The Engagement of Science and Religion* (Edinburgh: T&T Clark, 1998), 109.

cast as a clash of ideas, cosmological and theological, it is preeminently a clash between individuals. As has been pointed out, strictly speaking, ideas of themselves cannot clash; only people can.[12] And when people are involved, personality, personal interests, particular circumstances, passions—and so much more—play an enormous role. History is always about the particular and the human, and the particular and the human are always messy. The aim of this book is neither polemical nor apologetic. It seeks neither heroes and villains nor inevitable results based on impersonal social, economic, and cultural forces. Nothing is inevitable in human affairs unless people make it so. It is my hope that this book will, standing on the shoulders of so many others, help elucidate that. As it has been justly said, the past *is* another country, and it too needs its guides to be understood properly. The philosopher of science Ernan McMullin has stated that "had Galileo made his case for Copernicanism a century earlier or a century later, it seems unlikely that it would have evoked the strong response it did on the part of the Roman theologians."[13] One of the things to be gained by a better knowledge of the context is why that is so.

In the first chapter we cover the political background of Italy in Galileo's day, looking especially at the Grand Duchy of Tuscany, the Venetian Republic, and the Papal States, with Rome as its center. We will also discuss the religious situation of Italy and Counter-Reformation Catholicism with its organs of control, the Roman Inquisition, and the Index of Forbidden Books. In the second chapter we describe Galileo's life up to his discoveries with the telescope, the publication of the *Starry Messenger*, and his triumphant visit to Rome, including the humanistic and philosophical culture of his time. In the third chapter we discuss the history of science, particularly the history of astronomy and its often overlooked sister science of astrology, the historical relationship of Christianity to science and secular learning, and science in Galileo's day. In the fourth chapter we

[12] David C. Lindberg, "Galileo, the Church, and the Cosmos", in *When Science & Christianity Meet*, eds. David C. Lindberg and Robert L. Numbers (Chicago: University of Chicago Press, 2003), 34.

[13] Ernan McMullin, "Galileo on Science and Scripture", in *The Cambridge Companion to Galileo*, ed. Peter Machamer (Cambridge: Cambridge University, 1998), 274.

discuss Galileo's attempts to deal with his critics, primarily on the theological side, the events that led up to the 1616 condemnation of Copernicanism, and the election of Galileo's friend Maffeo Barberini as Pope Urban VIII. In the fifth chapter we cover the publication of his great work on cosmology, *Dialogue on the Two Chief World Systems* (1632), Galileo's trial before the Inquisition in 1633, and its aftermath. Finally, in the conclusion we will draw together various thoughts about the case and attempt a final reflection on it all.

During the Venetian Interdict crisis of 1606–1607 (which we will hear more of later), Cardinal Bellarmine, the defender of papal prerogatives, charged his Venetian opponents with "heresy in history" for saying that Charlemagne did not receive his empire as a gift of the pope. While profoundly revealing of the mentality of the papal side where, to use the phrase of Cardinal Manning after the First Vatican Council (1870) and the definition of Papal Infallibility, we see "the triumph of dogma over history", it does not negate the fact that short of divine revelation, there can be no heresy in history, only fallible human judgments about often barely understood human actions based on whatever often fragmentary evidence survives. Here, we manifest not the theological virtues of faith, hope, and charity, but the more mundane and profane virtues of honesty, carefulness, accuracy, and balance. We will see where it leads us. It should be an interesting ride.

Chapter One

Felix Italia: Italy in the Age of Galileo

In our textbooks and encyclopedias, we read that Galileo Galilei was born in Italy, and so we call him an Italian—and so we should. But we forget the *campanilismo* of Italian history: that profound local patriotism, that intense devotion to one's place, which even now marks the Italian character so many years after the peninsula's unification. A *toscano* is not a *siciliano*; and a *fiorentino* is not a *napoletano* or a *genovese*; and so on. Galileo, above all, was a Florentine and an inheritor of Florence's great intellectual, cultural, and artistic heritage, with Tuscan, the language of Dante, Petrarch, and Boccaccio, as its literary lodestar. While born in Pisa on February 15, he always proudly called himself a Florentine. During his time as a professor in Padua, he was listed on its rolls as a Florentine; on the cover of his *Starry Messenger* he is called *patritio fiorentino* ("a patrician Florentine"), and on the cover of the *Assayer* he describes himself as being a *nobile fiorentino* ("a noble Florentine"). It was the place of his dreams and the place to which he always returned. In trying to place Galileo and his family, the Florentine historian Giorgio Spini thought the best analogy would be to the great Brahmin families of Boston, seeing him as not unlike a Norton or a Prescott, if not a Winthrop or an Adams. As they were characterized and conditioned by their birth into the well-defined social and cultural tradition of Yankee New England, so was Galileo defined and conditioned by his birth into a well-defined social and cultural tradition—the tradition of the upper-class Florentine *cittadini* who formed its civic elite. In Spini's mind, Galileo was typical of them in his manly realism and in his love of the active life, and of anything useful and concretely practical; in his irony and in his high opinion and masterly use of the common tongue; in his enthusiasm for poetry, music, and the fine arts in general, and for Dante and

Michelangelo in particular; in his strong humanistic background, in his fondness for Plato and in his exalted conception of man's intellectual dignity; in his dedication to his family; and even in his genuine, if slightly anticlerical, religiosity.[1]

In the sixteenth and seventeenth centuries, Italy was just a geographical expression broken into many smaller states, with the most significant power being Spanish, not Italian. The Peace of Cateau-Cambrésis (1559) ended the constant fighting that Italy experienced since the French invasion of 1494 and left the Spanish Habsburgs controlling all of southern Italy (the Kingdom of Naples), Sicily, Sardinia, and the northern Duchy of Milan, as well as five garrisoned ports on the Tuscan coast. Half of Italy was under direct Spanish control. And with the descent of France into civil and religious war in the later sixteenth century, Spanish domination became ever more secure. The *Pax Hispanica* brought Italy fifty years of peace, and with peace, a revival of Italian prosperity. While historians of the *Risorgimento* (Italian unification), and their *bien pensant* fellow travelers, saw these years, the years of Galileo's life, as one of foreign domination and decay (artistic, intellectual, and spiritual), the people of that time saw it very differently—as one of the greatest, if not the greatest, periods in the whole of Italian history, and one in which Italian culture reached new heights.[2] Italian decline—and it was always a decline somewhat relative to the greater rise of the north Atlantic states—was not to begin until the early seventeenth century, and certainly from the 1620s, with the plague of 1630, which killed a third or more of the population in northern Italy, marking a poignant milestone of it.[3]

Italy was important to Spain not only as a source of revenue and military recruits and for maintaining her power in the Mediterranean, but also for connecting her to her Austrian Habsburg cousins in Germany and as part of the so-called Spanish Road, a safer overland

[1] Giorgio Spini, "The Rationale of Galileo's Religiousness", in *Galileo Reappraised*, ed. Carlo L. Golino (Berkeley: University of California Press, 1966), 44–66.

[2] Eric Cochrane, "Counter Reformation or Tridentine Reformation? Italy in the Age of Carlo Borromeo", in *San Carlo Borromeo: Catholic Reform and Ecclesiastical Politics in the Second Half of the Sixteenth Century*, ed. John M. Headley and John B. Tomaro (Washington, D.C.: Folger Books, The Folger Shakespeare Library, 1988), 31.

[3] Domenico Sella, *Italy in the Seventeenth Century* (London: Addison Wesley Longman, 1997), 19–49.

route to her possessions in the Low Countries during her many years of war against the Dutch. Between 1567 and 1620 more than 123,000 soldiers would travel it in the struggle to maintain the great Spanish empire.[4] Spain would remain the dominant power in Italy through Galileo's lifetime. So overwhelming did it seem that an Italian saying of the time remarked that "God has turned into a Spaniard." Italy did, however, retain her cultural preeminence. Her literature, music, and art were the model and admiration of Europe and many travelled to Italy to learn from her; her universities attracted thousands of foreign students; and Italy remained an essential stop for the sophisticated traveler, the devout pilgrim, and the callow youth seeking a bit of polish.

Amazingly enough for a man whose mind roamed the stars and who evinced such a deep vein of curiosity, Galileo seemed remarkably uninterested in other places and cultures. He never travelled outside Italy. In fact, he never travelled beyond a 170-mile radius from Florence, with Genoa to the north, Venice to the east, and Rome to the south as the extent of his range.[5] He never travelled to Naples, for instance, which at a quarter of a million people was not only the largest city in Italy but also very likely the largest city in all of Christian Europe and itself a vibrant center of intellectual life. Of the many Italian states, only three played an important part in Galileo's life: the Grand Duchy of Tuscany, the Republic of Venice, and the Papal States.

Galileo was born a subject of Duke Cosimo I (1519–1574), a member of the great Medici family to whom he would owe so much. While the Medici had reached some degree of eminence in Florence in the fourteenth century with Salvestro de' Medici being selected as *gonfalonier* in 1378, where he played a major role during the revolt of the unguilded cloth workers (the *ciompi*) and would be exiled in 1382, it was to the lesser branch of the family, of Salvestro's more modest cousin, Giovanni di Bicci de' Medici (c. 1360–1429), that the greatest of the Medici would spring.[6] Giovanni founded a bank in 1397, and when he died he was one of the richest men in

[4] Peter H. Wilson, *The Thirty Years War: Europe's Tragedy* (Cambridge, Mass.: Harvard University Press, 2009), 152.

[5] David Wootton, *Galileo: Watcher of the Skies* (New Haven, Conn.: Yale University Press, 2010), 14.

[6] Gene A. Bruckner, "The Medici in the Fourteenth Century", *Speculum* 32, no. 1 (January 1957): 1–26.

Florence.[7] His son Cosimo the Elder (1389–1464) would lead the bank to new heights, becoming in the process not only a great patron of the arts and learning but also the puppet master and unofficial head of the Florentine Republic. This political and artistic role continued in the family, reaching its apogee under Cosimo's grandson Lorenzo the Magnificent (1449–1492). While the Medici were thrown out with the invasion of the French in 1494, they were restored in 1512, only to be expelled in 1527 with the Sack of Rome at the time of the Medici Pope Clement VII (r. 1523–1534). Finally restored after an eleven-month siege by the Imperial army in 1530, they were to maintain their power over the city until the end of the dynasty in 1737.

Florence was a city perennially rent by factions within and coveted by hungry powers without, so it took great skill for the Medici to maintain their control, and a single misstep could be their last. As Eric Cochrane has pointed out, Italy in the sixteenth century was not yet a system of states as a system of big families, each engaged in a desperate game of musical chairs to seize one of few available states of the peninsula whenever the music stopped.[8] Some had done well for themselves (the Gonzaga of Mantua and the Este of Ferrara and Modena), some had fallen completely (the Sforza of Milan and the Borgia in the Romagna), and some were just entering the game, such as the Farnese under Pope Paul III (r. 1534–1549), whose illegitimate son Pier Luigi received the papal territories of Castro, Parma, and Piacenza with the title of duke.

Cosimo I de' Medici, a distant cousin of the main branch of the family, was made Duke of Florence in 1537 at eighteen (after the assassination of his profligate cousin Duke Alessandro) by the local aristocratic families who thought they could control him and by Emperor Charles V, whose troops had restored the Medici and still controlled the city. Secretive, severe, ruthless, and brusque in manner, he skillfully freed himself from the control of others, including Spain (though he remained a Spanish ally and married the daughter of the Spanish viceroy of Naples), and by thoroughness and discipline,

---

[7] Raymond De Roover, *The Rise and Decline of the Medici Bank, 1397–1494* (Cambridge, Mass.: Harvard University Press, 1963).

[8] Eric Cochrane, *Florence: The Forgotten Centuries 1527–1800* (Chicago: University of Chicago Press, 1973), 45.

built a strong, modern, centralized, absolutist, bureaucratic state. Under him Florence regained much of her splendor and prestige, gaining a degree of peace, freedom, and stability that had eluded her for a very long time. He patronized the sculptor Benvenuto Cellini (1500–1571) and after him Giorgio Vasari (1511–1574), famous for his *Lives* of the great Italian artists. He built the Uffizi Palace for his judicial and administrative offices. He doubled the size of the Pitti Palace, which became the residence of Cosimo and his family after 1550, and created the extensive Boboli Gardens behind it. Wishing to create a new Medicean Golden Age as splendid as the old, he labored and spent lavishly to make his court the equal of any in Europe in its intellectual and artistic eminence. He created the *Accademia Fiorentina*, the Florentine Academy (originally the *Accademia degli Umidi*), in 1540 to study and champion the Tuscan vernacular and, with the aid of Vasari, the prestigious *Accademia del Disegno* (the Academy of Drawing) in 1563, the first academy of art in Europe, which not only formed a kind of guild for Florence's artists but also taught painters, sculptors, architects, engineers, and designers of military fortifications.

Disegno, Italian for design or drawing, was a complex term associated with the tradition of Florence which referred not just to the capacity to render a form graphically, to the emphasis on line and the use of sketching and preparatory drawings as a building block to a composition, but also to the knowledge of the universal form from which the drawing was derived. It included a mastery of composition, anatomy, perspective, and proportion. It was contrasted with the Venetian emphasis on *colorito* "coloring", the judicious and direct application of color to the canvas in a spontaneous way, building up the painting as one went along. For Vasari, *disegno* was the guiding principle of the creative process and was as essential to sculpture and architecture as to painting. The *Accademia del Disegno* employed a professional mathematician to teach its students Euclidian geometry and perspective, and in 1589 Galileo would apply for the post though he would not receive it. Galileo's disciple and first biographer, Vincenzo Viviani (1622–1703), wrote that as a young man Galileo was so accomplished at drawing and so enjoyed it that if he could have chosen his own profession at that age he absolutely would have chosen to be a painter. His artistic knowledge was certainly respected by other artists. The most eminent Florentine artist of the day, Ludovico Cigoli (1559–1613)—who admitted that he learned perspective from Galileo—was his dear friend. Even if Galileo did not become a painter, he was formed in the spirit of Florentine *disegno*, and this brought a unique geometrical sensibility to his way of seeing the world. In 1613, however, he would enter the *Accademia del Disegno*: not as a student or teacher, but as an honored academician.

By the end of Cosimo's reign, the population of Florence had grown to some 80,000 from barely more than 50,000 in 1540. But it was still a far cry from the 120,000 it had in the 1340s before the Black Death of 1348 killed about half the population, or even the 90,000 of the 1480s. Not until the nineteenth century did Florence reach again its pre-plague population. In 1406 Florence absorbed the once great maritime Republic of Pisa, though Pisa would remain restive, regaining its freedom in 1494 and losing it again in 1509 after years of warfare and siege. Under Cosimo, whose court spent three months of every year there and who restored its university, Pisa became reconciled to its subjection to Florence and the city regained some of its glory. In 1557 Cosimo conquered the Republic of Siena to the south, unifying all Tuscany under his control, being invested with the title of Duke of Siena by Philip II the same year. As part of that investiture the Medici were obliged to supply the *soccorso* ("aid") of four hundred horsemen and four thousand infantry, a rather substantial force, anytime the Duchy of Milan or Kingdom of Naples were attacked—permanently tying the Medici to Spain. In 1569, to solidify further his position and raise him above the rest of the Italian princes, he gained the title of Grand Duke of Tuscany from Pope Pius V, which was confirmed by the emperor in 1576.

Cosimo's son Francesco I (1541–1587), taciturn and withdrawn, was more interested in his hobbies and collections, in his alchemy and curiosities, than in running the government, which he delegated to his ministers. During his reign, however, the arts continued to flourish in Florence, and the prestigious *Accademia della Crusca* (1583), dedicated to maintaining the purity of the Italian language and compiling a Tuscan dictionary, was created. While he produced seven children by Joanna of Austria, daughter of the Holy Roman Emperor Ferdinand I, only two daughters survived him, one being Marie de' Medici, who married King Henry IV of France in 1600. Francesco had a son, Don Antonio (1576–1621), by his Venetian mistress, Bianca Cappello, whom he legitimized after the death of Joanna (in 1578) and Francesco's marriage to Bianca. Francesco intended him to succeed him, but his brother Ferdinando moved quickly after Francesco's sudden and unexpected death to take the throne. Bianca, whom Ferdinando despised, had conveniently died within hours of her husband; and, while there were rumors of poison, the official report claimed malaria as the cause of death for both. While malaria has been found in Francesco's remains (it was endemic in the region and could be deadly—his mother and two brothers died of it in 1562), there is very strong evidence that the actual cause of death was arsenic, the preferred poison of the Renaissance.[9]

Don Antonio (1576–1621), only eleven at the time of his parents' death, joined the Knights of Saint John (Knights of Malta) when he came of age, and despite his vow of celibacy would father four children (whom he later had legitimized). While pushed aside by his uncle, he continued to live and serve at court, and he continued his father's patronage and passion for science. He was a correspondent and friend of Galileo (Galileo's discovery of scaling laws, for example, was announced in a letter to Antonio in 1609), and he will reappear later as a supporter of Galileo at the important breakfast discussion with the Grand Ducal family that sparked Galileo's *Letter to Castelli*.[10]

Ferdinando I (1549–1609), fifth son of Cosimo I and appointed a cardinal in 1563 at the age of thirteen, having learned the ways of the world as the head of the Medici faction at the papal court, resigned his ecclesiastical dignities to continue the family line, and under his skillful hands his realm and his family's wealth and prestige increased.[11]

---

[9] Francesco Mari, Aldo Polettini, Donatella Lippi, and Elisabetta Bertol, "The Mysterious Death of Francesco I de' Medici and Bianca Cappello: An Arsenic Murder?", *British Medical Journal* 333 (2006): 1299–301, and Donatella Lippi, "Still About Francesco de' Medici's Poisoning (1587)", letter to the editor, *American Journal of Medicine* 128, no. 10, October 2015, e61, http://www.amjmed.com/article/S0002-9343(15)00365-4/fulltext.

[10] Jacqueline Marie Musacchio, "Antonio de' Medici and the Casino at San Marco in Florence", in John Jeffries Martin, ed., *The Renaissance World* (New York: Routledge, 2007), 481–500.

[11] Elena Fasano Guarini, "'Rome, Workshop of All the Practices of the World': From the Letters of Cardinal Ferdinando de' Medici to Cosimo I and Francesco I", in *Court and Politics in Papal Rome 1492–1700*, ed. Gianvittorio Signoretto and Maria Antonietta Visceglia (Cambridge: Cambridge University Press, 2002).

He was affable, generous, just, and tolerant, genuinely desirous of ensuring the welfare and happiness of his people. He turned Livorno (Leghorn) into a major seaport. By offering extraordinary freedoms and privileges, including religious toleration, and building new harbor works and a canal connecting it to the Arno, the town rose from a little more than seven hundred at the beginning of his reign to just under six thousand at his death. He also spent heavily for other internal improvements and economic development, in land reclamation and irrigation, in navigational improvements, and in measures to increase trade. He even attempted to create a Tuscan colony in Brazil. Though he had been the cardinal protector of Spain, he was mistrustful of the Spanish, and as Grand Duke he tried to restore France as a serious counterweight to Spain. He gave substantial loans to support Henry of Navarre in his successful campaign for the French throne after the death of the last Valois king and worked to reconcile the papacy to him after his conversion by gaining for him papal absolution and recognition. After he was successful in this, he was able to arrange in 1600 the marriage of his own niece Marie to Henry. His own 1589 marriage to Christina of Lorraine, the favorite granddaughter of Catherine de' Medici, queen of France, produced nine children. It was to this same Christina, by then dowager Grand Duchess, that Galileo would address his famous discussion of science and biblical interpretation. Ferdinando's son Cosimo II (r. 1590–1621), whom Galileo tutored in the summers, would call Galileo back to Florence as his official philosopher and mathematician. Later on, when things became difficult for Galileo, the Medici would again be his strongest defenders and main support. While Florence under the Medici princes was certainly less creative when compared with its earlier history, it was equally far less turbulent and more externally magnificent.

> This city of Venice is a free city, a common home to all men, and it has never been subjugated by anyone, as have been all other cities. It was built by Christians, not voluntarily but out of fear, not by deliberate decision but from necessity. Moreover it was founded not by shepherds as Rome was, but by powerful and rich people, such as have ever been since that time, with their faith in Christ, an obstacle to barbarians and attackers.... This city, amidst the billowing waves of the sea, stands on the crest of the main, almost like a queen restraining its force.... As another writer has said, its name has achieved such

dignity and renown that it is fair to say Venice merits the title "Pillar of Italy," "deservedly it may be called the bosom of all Christendom." For it takes pride of place before all others, if I may say so, in prudence, fortitude, magnificence, benignity and clemency; everyone throughout the world testifies to this. To conclude, this city was built more by divine than human will.[12]

So wrote the Venetian patrician Marino Sanuto the Younger (1466–1536), the great chronicler of his home city in its heyday. Jacob Burckhardt in his classic *The Civilization of the Renaissance in Italy* (1860) compared Florence, "the city of incessant movement", with Venice, "the city of apparent stagnation", but which saw itself "as a strange and mysterious creation—the fruit of a higher power than human ingenuity".[13] The myth of Venice not only dominated Venetian life; it captured the imagination of outsiders who saw in the aristocratic republic a stability and social harmony so lacking in their own countries and idealized the Venetian constitution as a political model, as a paragon of liberty and order. Its Doge, Senate, and Great Council seemed to manifest the Aristotelian ideal of mixed government that combined the best elements, respectively, of monarchy, aristocracy, and democracy. Venice was the only Italian state free of foreign control and remained so until the time of Napoleon in 1797. While the Grand Duchy of Tuscany was Galileo's home, the most productive and the happiest years of his life (as he himself admitted) were not there, but in the Republic of Venice. From 1592 until 1610, Galileo taught at the University of Padua in Venetian territory, in *La Serenissima*, in the Most Serene Republic of Venice.

Sprung from several refugee lagoon communities, by 1400 this city in northeastern Italy, built on 117 small islands two and a half miles off the Italian mainland and a little less than two miles from the northern Adriatic, had become the greatest sea power in the Mediterranean. Her fleets of galleys sailed as far west as Portugal, England, and Flanders on the Atlantic, as well as to the south and east, to Egypt and the Levant in the Mediterranean. She had monopolized

---

[12] David Chambers and Brian Pullan, eds., *Venice: A Documentary History, 1450–1630* (Blackwells: London, 1992), 4–5.

[13] Jacob Burckhardt, *The Civilization of the Renaissance in Italy*, trans. S. G. C. Middlemore (New York: Random House, 1954), 51.

the Asian spice trade and possessed territory all across the Greek archipelago. In the early fifteenth century she expanded along the Dalmatian coast as far south as Albania and deep into the Italian mainland near her (the Terraferma) so as to protect her overland trade routes. In 1402 she conquered Verona, in 1405 Padua, and in the years following, the rest of eastern Lombardy. The Peace of Lodi (1454) ended these series of wars between the major northern Italian states and confirmed Venetian control of a swath of territory across northeastern Italy, from Friuli and the Veneto in the east to Brescia and Bergamo in the west. In 1508 the European-wide League of Cambrai was formed, the greatest grouping of states Europe had seen up to that time. It allied King Louis XII of France, Emperor Maximilian, King Ferdinand of Aragon, Pope Julius II, Mantua, and Ferrara, and sought to divide Venice's Italian empire among themselves. It nearly brought Venice to complete collapse. But she survived the crisis, and by 1517 she had recovered all her territories. In Galileo's day, caught between the Austrian Habsburgs to the north and east and Spanish Habsburgs to the west and south, and the ever-threating Turks to the east, she maintained a precarious independence.

In an age of increasing absolutism, Venice maintained its unique aristocratic and remarkably stable republican structure. According to traditional Venetian historiography, the year 1297 was seen as the key date in Venice's constitutional history. This was the so-called *Serrata*, or closing of the Great Council, when membership there was limited to about 150 families, or about twelve hundred adult males. It now seems that the process of settling the noble class took some decades longer, until 1323, with new families being added after the War of Chioggia against Genoa (1379–1381), and no other families being added, with few exceptions, until the exhausting Cretan War (1645–1669) against the Turks.[14] Founded on trade, long inured to service of the state, and formed by its practical, sober, prudential ethos, these members of the Great Council, these nobles or patricians, adult males

---

[14] Gerhard Rösch, "The Serrata of the Great Council and Venetian Society, 1286–1323", in *Venice Reconsidered: The History and Civilization of an Italian City-State 1297–1797*, ed. John Jeffries Martin and Dennis Romano (Baltimore: Johns Hopkins University Press, 2002), 67–88.

older than twenty-five (except for clerics), formed the active citizenry of the nation and filled all major government offices—perhaps four hundred to five hundred nobles holding offices at any one time, though for an efficiently run government only about one hundred men of wealth, ability, and public spirit were needed for its principal offices. To ensure a sufficient supply of qualified officials, it even penalized those who refused to take office. In the fifteenth and sixteenth centuries, rules were developed governing marriages with commoners, examining claims of nobility, and excluding illegitimate sons from noble rights—all to preserve the integrity of the noble body.[15] And while it is all too easy to idealize the Venetian patriciate and to ignore its deficiencies, as many historians have, still in the sixteenth century there was nothing like it among the nobility of Europe, with an attachment to tradition and a cohesion as a ruling class that was remarkable. In the early sixteenth century, the number of adult male nobles was slightly above twenty-five hundred, and noble families formed about 6 or 7 percent of the city's population. In the 1570s they lost about a quarter of their members from the Battle of Lepanto (1571) and the plague of 1575–1577 (the same plague that reduced the population of Venice from 170,000 to 120,000), but they recovered somewhat over the next few decades, though never to reach their peak numbers again. The numerical decline of the nobility was accelerated by the restriction of marriage among its members as their economic opportunities declined, and as dowry inflation and the need to preserve family patrimony demanded the sacrifices of individual members from both genders. Nearly half of patrician men remained bachelors (some entering the Church as clerics), and between 1581 and 1642 over 70 percent of patrician women entered convents.[16]

As necessary for the proper functioning and the stability of the Venetian state as its patricians were, the *cittadini*, its elite citizens, another 5 to 8 percent of the population, some 208 families by 1540,

---

[15] Stanley Chojnacki, "Identity and Ideology in Renaissance Venice: The Third *Serrata*", in Martin and Romano, *Venice Reconsidered*, 263–94.

[16] William J. Bouwsma, *Venice and the Defense of Republican Liberty* (Berkeley: University of California Press, 1968), 60; James Cushman Davis, *The Decline of the Venetian Nobility as a Ruling Class* (Baltimore: Johns Hopkins University Press, 1962), 17–25, 54–74; and Gianna Pomata, "Family and Gender", in *Early Modern Italy 1550–1796*, ed. James Marino (Oxford: Oxford University Press, 2002), 78.

were also essential. They had the right of entry into the highest offices of the state bureaucracy, the leadership of the greater confraternities (*Scuole Grandi*) which were so important to Venetian religious, charitable, and artistic life, and they possessed special trading privileges. Many of these families had a blood relationship to noble families, past or present, and altogether they formed a kind of "cadet nobility" who often possessed the wealth and skills lacking in so many of the patricians.[17]

With a population of 170,000 (there was an additional two million on her mainland possessions), Venice was the most exotic and cosmopolitan city in Europe, as well as one of the richest and most magnificent. Here one jostled with tourists and merchants, with pilgrims to the Holy Land and prostitutes, with Germans and Greeks, Armenians and Jews, Slavs and Albanians, and even the Turks had a place to call their own. Foreigners were sometimes appalled at the intermingling of nations and classes that they saw in Venice. The Venetian Arsenal, described by Dante in his *Inferno*, and to which Galileo would become a consultant in 1593, covered about 110 acres or 15 percent of the city, was a complex of state-owned shipyards and armories, and the greatest industrial enterprise in Europe. Employing some sixteen thousand workers at its peak in the early sixteenth century, and using an almost assembly-line process and prefabricated, standardized, interchangeable parts, it could build and outfit a fully equipped ship within a day. Besides the outfitting of ships, it also maintained them, and in the Arsenal was kept a reserve of ships and armaments to assemble an armed fleet speedily if necessary. More than half of the ships at the great Battle of Lepanto were Venetian built, and without Venetian support the victory would not have happened.

The Battle of Lepanto (October 7, 1571), fought when Galileo was just a child, was one of the greatest naval engagements in history—the last great galley battle—and bulked large in the consciousness of sixteenth-century Europe. The threat of Turkish conquest of Europe intensified in the sixteenth century when in 1526 the Turks under Sultan Suleiman the Magnificent annihilated the army of King Louis II of Hungary and Bohemia at the Battle of Mohács, and in 1529 he besieged Vienna, with perhaps one hundred thousand men.

---

[17] James Grubb, "Elite Citizens", in Martin and Romano, *Venice Reconsidered*, 339–64.

On the sea, the Turkish fleets and the constant raids by Muslim corsairs made travel in the Mediterranean Sea or living on its coast dangerous things. (One recent scholar has estimated that at least a million Europeans were enslaved by the Barbary pirates between 1530 and 1780.[18]) In the second half of the sixteenth century the Turkish threat took on an ever-greater urgency as the Turkish rule was extended all the way to Morocco in North Africa and they attacked the Knights of Saint John on the island of Malta in 1565 (heroically beaten back with Spanish aid) and conquered in 1570–1571 Cyprus, Venice's richest possession in the eastern Mediterranean. A Holy League was formed between Spain, Venice, Genoa, the papacy, and some lesser states, and on October 7 the Christian fleet of 212 ships led by Don John of Austria (younger half brother of Philip II) crushed the 286 ship Turkish fleet led by Ali Pasha, capturing his flagship and killing him in the process. The Turks lost 210 ships (of which 130 were captured) and suffered 25,000 casualties while the Christians lost only 50 ships and 13,000 casualties. In a matter of just four or five hours, in the Gulf of Corinth in Greece, in the bay of Lepanto, the Turkish menace seemed to go up in smoke.

While the victory at Lepanto certainly gave the Christian powers an incredible psychological lift, saved Italy from certain invasion, and produced a prodigious quantity of art, literature, and music celebrating it, and while it may have destroyed, as some historians believe, the myth of Turkish invincibility, it was not a true turning point.[19] As the Turkish sultan, Selim II, was said to have remarked, "The infidels only singed my beard; it will grow again."[20] Within two years the Turks rebuilt and restaffed their fleet, and their expansion continued, forcing Venice to arrange a separate truce in 1573. While Venice was much criticized for her abandonment of the war, in 1580 even militant Spain had to arrange a truce with the Turks—and not on the best terms. The Turks would remain a threat into the eighteenth century. Lepanto, however, did inspire a new spirit and a new pride in

[18] Robert Davis, *Christian Slaves, Moslem Masters: White Slavery in the Mediterranean, the Barbary Coast, and Italy, 1500–1800* (New York: Palgrave Macmillan, 2004).

[19] Andrew C. Hess, "The Battle of Lepanto and Its Place in Mediterranean History", *Past & Present* 57 (1972): 53–73.

[20] Carlo M. Cipolla, *Guns and Sails in the Early Phase of European Expansion, 1400–1700* (London: Collins, 1965), 100.

Venice. It also made her aware of the many dangers she had to navigate, dangers posed by the Habsburgs, by the Turks, and by a newly aggressive papacy, and also of the dangers of drift and inaction that were leading to a loss of Venice's maritime commerce and her mercantile virtues. This awareness found its outlet in a younger group of the Venetian ruling class, the *giovani* (the young), who overthrew in the winter of 1582–1583 a small group of very powerful and wealthy families who had long dominated the affairs of the state. Many of Galileo's Venetian friends would be among this group.

> Not only is there bread in abundance; there is also an incalculable wealth of all goods and delicacies, which are brought hither, not only by the rivers and canals of the mainland, but also by the sea, from as far afield as Egypt, Syria, the Archipelago, Constantinople and the Black Sea. To Venice come the oils of Apulia, the saffrons of the Abruzzo, the malmseys of Crete, the raisins of Zante, the cinnamon and pepper of the Indies, the carpets of Alexandria, the sugar of Cyprus, the dates of Palestine, the silk, wax and ashes of Syria, the cordovans of the Morea, the leathers, *moronelle*, and caviar of Caffa. There is such a variety of things here, pertaining both to man's well-being and to his pleasure, that, just as Italy is a compendium of all Europe, because all the things scattered through the other parts are happily concentrated in her, even so Venice may be called a summary of the universe, because there is nothing originating in any far-off country but it is found in abundance in this city. The Arabs say that, if the world were a ring, then Ormuz, by reason of the immeasurable wealth that is brought thither from every quarter, would be the jewel in it. The same can be said of Venice, but with much greater truth, for she not only equals Ormuz in the variety of all merchandise and the plenty of all goods, but surpasses her in the splendor of her buildings, in the extent of her empire, and, indeed in everything else that derives from the industry and providence of men.[21]

In the description of Giovanni Botero (1544–1617) above, we see well Venice's pre-eminence as a trading center. Venice was built on trade, had grown rich on trade—trade of every sort, making it the greatest emporium in all Europe. And a great part of her success

---

[21] Chambers and Pullan, *Venice: A Documentary History*, 167–68.

had been due to her monopoly of the spice trade into Europe. But when in 1498 Vasco da Gama found another route to the Indies around Africa's southern cape, one might have thought that Venice's days were numbered, especially after the Portuguese captured the trade's main Asian ports. However, such was not the case. The Portuguese were not able to cut off completely, or for too long, the flow of spices through the Levant, and they sought to keep prices up so that Venetian spices were not pushed out of the market. By the mid-sixteenth century Venice had recaptured much of the trade. Nor was she dependent upon it so that even in the periods when she was totally cut off, she had other means of success. The Levant and Egypt remained the main source of her raw materials like cotton and a major market for her manufactures. Rather, other factors were far more destructive of her mercantile empire: the on-again, off-again wars with the Turks, which would continue into the seventeenth century and later; the Thirty Years' War (1618–1648), which disrupted her German trade; the Dutch capture of the Spice Islands in the early seventeenth century; and the movement of the Dutch and English into the Mediterranean in the same century.

While the Venetian Republic could be as brutal and ruthless as any state when it felt threatened, and act with a speed and secrecy that surprised foreign observers, it was famous for its great tolerance. When compared with most European states, it allowed, in some ways, an extraordinary amount of freedom. William Thomas (c. 1507–1554)—a Welsh civil servant who fled from justice to Italy in the late 1540s, having pocketed the money of his patron, an important member of court, and who would later become Clerk of the Privy Council and intimate of the young King Edward VI—was surprised by the degree of freedom there, particularly for foreigners.

> All men, specially strangers, have so much liberty there that though they speak very ill by the Venetians, so they attempt nothing in effect against their state, no man shall control them for it. And in their *Carnevale* time (which we call Shrovetide) you shall see maskers disguise themselves in the Venetians' habit and come unto their own noses in derision of their customs, their habit, and misery.
>
> Further, he that dwelleth in Venice may reckon himself exempt from subjection. For no man there marketh another's doings, or that meddleth with another man's living. If thou be a papist, there shalt

thou want no kind of superstition to feed upon. If thou be a gospeler
[Protestant], no man shall ask why thou comest not to church. If thou
be a Jew, a Turk, or believest in the devil (so thou spread not thine
opinion abroad), thou art free from all controlment. To live married or
unmarried, no man shall ask thee why. For eating of flesh in thine
own house, what day soever it be, it maketh no matter. And generally of all other things, so thou offend no man privately, no man shall
offend thee, which undoubtedly is one principal cause that draweth so
many strangers thither.[22]

In the ever more acrimonious period of the Reformation and
Counter-Reformation, Venice was an anomaly. In 1547, under pressure from Rome, the Venetians did create the *Tre savi all'eresia* to
deal with heresy and function as a branch of the Roman Inquisition
which had been instituted in 1542 by Pope Paul III; but this new tribunal would consist not only of three clerics (the patriarch, the local
inquisitor, and the papal ambassador), but also of three noble laymen chosen by the government. While the Venetians had no love
of heresy and saw themselves as a particularly Catholic nation, they
were also desirous of preserving their independence and maintaining
friendly trade relations with non-Catholic nations. They were also
suspicious of some of the more extreme views showing themselves in
the Eternal City. Some twenty-five heretics were executed in Venice
during the sixteenth century.[23] Despite the urging of Rome for some
more dramatic public spectacle, they were drowned in secret at night
before dawn in the Adriatic. This was partially to protect Venice's
reputation for tolerance and partly to cause fear.

The Venetian Church was remarkably independent of Roman
control. The patriarch, the bishops, and other higher clergy were
voted on by the Senate before the names were sent to Rome for
confirmation; parish priests were elected by the taxpayers of the parish; Church property was taxed; and accused clerics were judged by
secular courts. Within Venice itself its Greek Orthodox subjects had
their own church, San Giorgio dei Greci, with its concomitant educational, cultural, and and publishing institutions; Jews and Muslims

[22] William Thomas, *The History of Italy* (Ithaca, N.Y.: Cornell University Press, 1963), 83.
[23] John Jeffries Martin, *Venice's Hidden Enemies: Italian Heretics in a Renaissance City* (Baltimore: Johns Hopkins University Press, 1993), 69.

could perform their religious rituals in their own quarters or hostels; and foreign Protestants, if they did not proselytize, propagandize, or cause a ruckus, were left alone. The University of Padua attracted so many students from across Europe not only because it was perhaps the finest university in Europe but also because it did not demand a Catholic confession of faith on the reception of degrees—something that was not true of other Italian universities.

> All those who devote themselves to the composition of new works, or the restoration and correction of ancient ones, not only for their own benefit but for that of others (for, as Plato has wisely said, we are not born for ourselves alone, but partly for our native land, partly for our parents, and partly for our friends), all those I say, need peace and quiet, and betake themselves from the concourse and company of men into solitude, as into a harbor. For the sacred studies of letters and the Muses themselves always require leisure and solitude.... But as for me, there are two things especially, not to mention some six hundred others, which interrupt and hinder my zealous studies.... And so that those who come to say "hello," or for any other reason, may not continue to interrupt my work and serious study, I have taken care to warn them, by putting up a notice, like an edict, on the door of my office to the effect: WHOEVER YOU ARE, ALDUS BEGS YOU ONCE AND FOR ALL TO STATE BRIEFLY WHAT YOU WANT, AND THEN LEAVE QUICKLY, UNLESS YOU HAVE COME, LIKE HERCULES, TO SUPPORT THE WEARY ATLAS ON YOUR SHOULDERS, FOR THAT IS WHAT YOU WILL DO WHEN YOU ENTER THIS WORKSHOP.[24]

Aldus Manutius (1440–1515) was one of the great publishers, if not the greatest publisher, of the Renaissance, his famous anchor-and-dolphin logo even being picked up by later publishers, including Doubleday at present. In the fifteenth and sixteenth centuries, Venice—with its mercantile connections, its cosmopolitan and polyglot population, its well-developed diplomatic and intelligence networks, and its strategic location—was the leading center not only for the gathering and dissemination of information in Europe, but also for the book trade, of which the Aldine Press of the House of Manutius

[24]James Bruce Ross and Mary Martin McLaughlin, eds., *The Portable Renaissance Reader* (New York: Viking Press, 1953), 396–98.

was its greatest ornament. In 1469 Johan von Speyer, Venice's first printer, arrived in the city and inaugurated the Venetian publishing industry. By 1500 about twenty-five German printing firms were in the city plus many others, both from abroad and from other parts of Italy, including Aldus Manutius, who was from Bassiano to the south of Rome and settled in Venice in 1490. Manutius, a humanist and friend of the Renaissance polymath Giovanni Pico della Mirandola, educated in Latin by Gasperino da Verona at Rome and in Greek at Ferrara by the famous Guarino da Verona, gathered around him Greek scholars and workers to produce important editions of Greek classical works such as Aristophanes, Sophocles, Euripides, Plutarch, Plato, Pindar, Galen, Thucydides, Herodotus, Demosthenes, and Aristotle. He also published editions of Latin and Italian classics. Among his more contemporary authors were Erasmus in his *Adagia* and Castiglione in his *Book of the Courtier*. He strove not only for superb typography and book design (he introduced, for example, the use of italic print) but also for relatively inexpensive, handy, elegant, pocket-sized editions that could be easily carried and easily read. He also introduced a standardized system of punctuation, with the first modern use of the comma and the first printed semicolon.

Hundreds of publishers made their home in Venice, which attracted many professional writers, a veritable "Venetian Grub Street".[25] It was also a center for publishing Hebrew works (e.g., the Talmud), non-classical Greek texts, works in the Slavonic languages and in Spanish, as well as translations into Italian, everything from the classics to works originally in Arabic and other European languages, particularly translations from the Spanish. Books of every sort and every taste were published in Venice, and so abundantly and cheaply that it put them within the reach even of many artisans. In 1596 the presses of Venice employed four hundred to five hundred men. By the middle of the sixteenth century Venice produced and sold more books than any other European city, and in the same century an estimated fifteen million to twenty million books were printed there.[26]

---

[25] Peter Burke, "Early Modern Venice as a Center of Information and Communication", in *Venice Reconsidered: The History and Civilization of an Italian City-State, 1297–1797*, ed. John Martin and Dennis Romano (Baltimore: Johns Hopkins University Press, 2000), 389–419.

[26] Paul Grendler, *The Roman Inquisition and the Venetian Press, 1540–1605* (Princeton, N.J.: Princeton University Press, 1977), 3, 12.

Galileo enjoyed Venetian intellectual freedom, both in his teaching at Padua and in his jaunts to Venice itself, about twenty-five miles from Padua, where he would frequent the salon of the Morosini brothers, Andrea and Niccolò, a center of the *giovani* and around whom many of the leading Venetian intellectuals and politicians could be found. Throughout the sixteenth and early seventeenth centuries, it was the constant complaint of Roman churchmen that Venice was a haven of heresy and infidelity. When Galileo was to leave Padua in 1610, his friends warned him that he might find life less congenial in a more clerically controlled Tuscany since the Venetians were also far more protective of their subjects from the curiosity and probings of the Inquisition.

Venice's relations with the papacy had always been mixed, but with the far more aggressive papacy of the later sixteenth century they took a turn for the worse—especially when Venice extended to her mainland possessions in 1602, 1603, and 1605 some of the laws that applied to the city in reference to the construction of new churches and alienation and leasing of lay property to the Church, and most particularly when it arrested and tried in a secular court two clerics for civil crimes. It was during Galileo's time at Padua, in 1606, that Pope Paul V excommunicated the Doge and the entire Venetian Senate, and placed an interdict on Venice which should have suspended all religious services and sacramental activity in the Republic—all marriages, baptisms, Masses, burial services, and so on—until Venice had been absolved. Having experienced this before in 1482 and 1509, the Serenissima challenged the legitimacy of the act, refused to have the interdict published in her territories, denied its effects, and waged a campaign internally and internationally against it, calling upon the services of Galileo's good friend the Servite friar Fra Paolo Sarpi (1552–1623) as a theological advisor, defender, and propagandist. Sarpi had served under Carlo Borromeo in Milan, and was later the provincial of his order in Venice and his order's procurator general or agent in Rome, where he came to know Bellarmine and the Roman Curia well. He returned to Venice, where he belonged to the Morosini circle. He was not only a master of history and canon law (his 1619 *History of the Council of Trent* would become a major thorn in the papacy's side), but a brilliant mathematician and scientist. Galileo called the much older man "my father and my master" and used him as his sounding board. It was Sarpi, for instance, who

introduced the telescope to Galileo. The English ambassador to Venice, Sir Henry Wotton (1568–1639), was greatly impressed by him.

> He seemeth, as in countenance so in spirit, liker to Philip Melanchton
> than to Luther, and peradventure a fitter instrument to overthrow the
> falsehood by degrees than on a sudden; which accordeth with a fre-
> quent saying of his own, that in these operations *non bisogna far salti*.
> He is by birth a Venetian, and well skilled in the humours of his own
> country. For learning, I think I may justly call him the most deep and
> general scholar of the world, and above other parts of knowledge, he
> seemeth to have looked very far into the subtleties of the Canonists,
> which part of skill gave him introduction into the Senate. His power
> of speech consisteth rather in the soundness of reason than in any other
> natural ability. He is much frequented, and much intelligenced of
> all things that pass; and, lastly, his life is the most irreprehensible and
> exemplar that hath ever been known. These are his parts, set down (I
> protest unto your Lordship) rather with modesty than excess.[27]

Sarpi is an enigma. Some believed him a Protestant in a friar's
cowl; some believed him a secret skeptic, a scoffer at traditional
Christianity; some believed him a Catholic of the old school. And
there is plenty of evidence for each. We will never know for certain.

Sarpi's argument was that Venice was defending an older and more
traditional Catholicism against more recent exaggerations of papal
prerogatives and that the pope had no right to intervene in the affairs
of a sovereign state whose power was given by God, a power certainly extending to jurisdiction over church property and the clerics
in its territory. So successful was Sarpi's writing that he was not only
excommunicated by the pope but he was also the object of an assassination attempt. On October 5, 1607, he was attacked by three men
who stabbed him many times, leaving three wounds (two in the neck
and one in the temple) before they fled to the safety of the Papal
States. He survived, but papal animosity did not end, even with his
death in 1623, for the papal nuncio to Venice then demanded that
his body be dug up and put on trial for heresy. The Serenissima did not
acquiesce in this ghoulish request. Even as late as 1722 the Inquisition

[27] Logan Pearsall Smith, *The Life and Letters of Sir Henry Wotton*, vol. 1 (1907; repr., Oxford: Clarendon Press, 1966), 400.

pursued him, trying to have Sarpi disinterred from his honored place of burial and relegated to a common dumping ground for the dead.[28]

> And because I am fallen in to the mention of the people, that you may at once understand the present face of Religion here.... Such a multitude of idolatrous statues, pictures, reliques in every corner, not of their churches only, but houses, chambers, shopps, yea the very streets, and in the country the high wayes and hedges swarme with them. The sea it self is not free; they are in the shipps, boats and water-marks. And as for their slavery and subjection to them, it is such, as that of paganisme came not to the half of it. Whereof to give such a taste as may be allso for some cause of it; noe sooner doe their Children almost creep out of their Cradles, but they are taught to be Idolators. They have certain childish processions, wherein are carried about certain puppets, made for their Lady, and some boy that is better Clerke than his fellows goes before them with the words of the Popish Litany; where the rest of the fry following make up the quire. A great tyrant is custome and a great advantage hath that discipline which is suck'd in with the mothers milke.[29]

As we see from this 1608 description of William Bedell, the chaplain to the English ambassador Sir Henry Wotton, Catholicism pervaded every aspect of Venetian life. While many northern Protestants believed that the Italian Reformation had finally begun— and Sir Henry Wotton certainly labored for one, even having his chaplain translate the English Anglican liturgy into Italian for their use—and that a full-scale war would break out with Venice allied to the northern Protestants against the Catholic power of Rome and the Habsburgs, there was never any possibility of Venice changing its religion, whatever the pope might do. Eventually, after a year (the interdict only lasted from May 1606 to April 1607), reconciliation was arranged by France, though in almost all respects it was a victory for the Venetians. Long after he left her, Venice was to remain in Galileo's thoughts. It is not surprising that his two greatest works, his *Dialogue on the Two Chief World Systems* (1632) and his *Discourses on Two New Sciences* (1638), were both set in Venice.

---

[28] Bouwsma, *Venice and the Defense of Republican Liberty*, 624 and n. 3.
[29] Chambers and Pullan, *Venice: A Documentary History*, 196.

Finally, there were the Papal States, or more precisely, Rome, the Eternal City, which Galileo would visit six times in his life. His first visit was in 1587, as a twenty-three-year-old suppliant seeking letters of support from major Roman figures to help him gain a university appointment. The final visit was in 1633, as a sick man of sixty-nine, soon to start going blind, for his trial and condemnation before the Inquisition. As the seat of the papacy, and therefore of the Catholic Church, it was the center of Catholic life, patronage, and scholarship, and a place where an ambitious Galileo would be inevitably drawn.

The Papal States occupied the center of Italy. With its powerful baronial families, Rome had always been difficult to control, and many medieval popes had been forced to flee the city. In 1309, the pope and his central bureaucracy, the Roman Curia, settled in Avignon, leaving Rome to her devices. In 1377 Pope Gregory XI returned, but his death the next year led to the election of Urban VI and the Great Schism with popes both at Avignon and Rome, and a Europe bitterly divided between the two. Eventually there were even three popes. In 1417, the Council of Constance (1414–1418) led to an end of the schism with the election of a new pope to replace the rest: Martin V, who returned to Rome in 1420. Deeply involved in Italian affairs, in its politics and in its wars, the popes expanded their temporal power beyond the territories around Rome until it extended across the Apennines into the Marches, Romagna, and even beyond. In 1598 the papal fief of the Duchy of Ferrara was annexed by force after the death of the last legitimate ruler of that branch of the Este, who had been lords of the city since the thirteenth century. And in 1625, another papal fief, the Duchy of Urbino, was annexed, bringing the papal boundaries to the form they would have until the nineteenth century.

The sixteenth century was to see Rome grow from a city of thirty thousand people to one of one hundred thousand, and from a medieval city still living in the ruins of a grand past into a cosmopolitan city, a center of Italian and world affairs, and a worthy capital of Christendom—though it was only in the seventeenth century that some of its most famous architectural and artistic monuments were finished. In the course of the century, the Papal States were transformed into a highly centralized bureaucratic state, equal if not superior to other highly centralized bureaucratic states of early modern

Europe—though in this case run by priests—and from one of the least taxed Italian states probably to one of the most taxed, by 1600 paying four times what they did in 1500. In the early fifteenth century the Papal States produced 25 percent of papal revenue; in 1521 they produced some 37 percent; and by the end of the sixteenth century some they accounted for 80 percent. To this increase in taxation, which was never sufficient for papal needs, was also added massive public debt, of papal bonds (*luoghi di monte*) floated all over Italy and beyond, which made Rome the center of high finance, one of the most secure money markets in Europe, and the haunt of Florentine and Genoese bankers. By 1599 more than half of the total papal income had to be used for the interest payments on these papal bonds. Like other monarchies of the time, the papal monarchy took on all the trappings of an absolutist court, a place where conspicuous consumption displayed and maintained one's status—and it was intensely status conscious—and where connections to the prince, the pope-king, meant everything.[30]

By 1600, there were some sixty-seven palaces and twenty villas in and around the city. Besides the papal court and its administration, there were also the *famiglia* or entourages of many cardinals (some fifty resided in Rome in 1603), of foreign ambassadors, and of the great noble families, with numerous hangers-on—and the wealthier the cardinal or more powerful the ambassador or grander the noble, the larger the entourage, and some were in the hundreds. It was expected that even the poorest cardinal would maintain a decent-sized and respectably appointed household; assistance was provided from papal funds for the poorer cardinals to do so. Numerous students attended the Roman universities, and many consecrated men and women filled its religious institutions. In the early seventeenth century there were between four thousand and six thousand courtiers and students in the city, and about five thousand clergy. And while there was plenty of construction work from the second half of the sixteenth

---

[30] Jean Delumeau, "Rome: Political and Administrative Centralization in the Papal State in the Sixteenth Century", in *The Late Italian Renaissance: 1525–1630*, ed. Eric Cochrane (London: Macmillan, 1970), 287–304; Peter Partner, "Papal Financial Policy in the Renaissance and Counter-Reformation", *Past & Present* 88 (1980): 17–62; and Paolo Prodi, *The Papal Prince: One Body and Two Souls; The Papal Monarchy in Early Modern Europe*, trans. Susan Haskins (Cambridge: Cambridge University Press, 1987).

century on (some fifty-five new churches were built, mostly after 1550, and many new palaces) and many artisans to produce high-end goods for the rich, there was little real industry. Despite its outward luster, it was a city of endemic poverty, a poverty made worse in the late sixteenth century, when famine repeatedly struck the land. It had a large floating population of beggars, pilgrims, and migrant day laborers, with the Jubilee Year of 1574–1575 bringing some 139,000 pilgrims to overload the city's charitable organizations. It was also an overwhelmingly male city and violence was common, with bandits roaming the outlying countryside (it was particularly bad between 1578 and 1595), and within the city itself where, despite the legal prohibitions against the carrying of weapons without a license, there were frequent armed fights and bloody brawls. Men outnumbered women by a significant margin, sometimes 150 or even 178 men to 100 women (a surplus of between twenty thousand and thirty thousand men in the early seventeenth century), with a large number of single men and about a thousand full-time prostitutes.[31] But as the French writer Montaigne recounts, visiting Rome in 1580, even the prostitutes reflected the piety of the holy city.

> While a certain man was in bed with a courtesan, amid the license of that relationship, behold at midnight the Ave Maria rang; she immediately jumped out of bed onto the floor and got on her knees to say her prayer. When a man was with another courtesan, behold the good mother (for the young ones especially have old governesses, whom they treat as mothers or aunts) comes banging at the door, and with anger and fury tears from this young one's neck a ribbon from which hung a little Madonna, so as not to contaminate it with the filth of her sin. The young one felt extreme contrition for having forgotten to take it from her neck as she was accustomed to do.[32]

The Rome of Galileo's time was preeminently the Rome of the Catholic Reformation, or as older historians preferred, the Counter-Reformation. While it never stopped being a city of ambitious clerics

---

[31] Gigliola Fragnito, "Cardinals' Courts in Sixteenth Century Rome", *Journal of Modern History* 65 (1993): 26–56; Tessa Storey, *Carnal Commerce in Counter-Reformation Rome* (Cambridge: Cambridge University Press, 2008), 57–60, 67, 106, 131.

[32] Michel Montaigne, *Travel Journal*, trans. Donald M. Frame (San Francisco: North Point Press, 1983), 85.

on the make and ecclesiastical networking, it also became the city of the amiable Saint Philip Neri, of the humble Capuchin lay brother Saint Felix de Cantalice (1515–1587), and of the newly reopened catacombs. The Renaissance papacy that preceded it had been a glorious era of artistic brilliance, and humanity owes a large measure of gratitude to the papacy for its patronage of men like Michelangelo (think of the great ceiling and the *Last Judgment* of the Sistine Chapel) and Raphael (think of the *Stanza della segnatura*), to name but two. And by 1500 the humanist men of letters had truly made Rome the *Caput Mundi*, the Head of the World, the unrivalled and undisputed center of literary and intellectual life in Europe.[33] But this cultural primacy came at a cost, and by draping itself in the advance guard of worldly taste, the papacy opened itself up to the eternal protest of the Gospel, whose Kingdom is not of this world, and to the revolution which was the Reformation. The easygoing and brilliant city of the Renaissance popes, of the era of the young Copernicus— with its fluid mix of paganism and Christianity, of personal laxity and formal piety—was no more, or at least not in such an obvious way. Thus, for example, while prostitution was still permitted and taxed, it was restricted to certain areas of the city and more strictly policed, and while Carnival and other "occasions of sin" continued to be allowed, they were given less prominence by Roman authorities—and popes were no longer such obvious participants.[34] No more the world of the Borgia pope, Alexander VI (r. 1492–1503), with his mistresses and children; no more the world of Julius II (r. 1503–1513), the warrior pope dressed in armor; no more the world of the pleasure-loving Pope Leo X (r. 1513–1521), with his buffoons and hunting dogs. The reign of Leo X, the son of Lorenzo the Magnificent, was greeted as the beginning of a new Golden Age, a time of peace and concord after the belligerent reign of Julius II. Personally chaste, moderately devout, a lover of the good things of life, generous and good-natured, Leo X was an exuberant patron to scholars and artists, but his spendthrift ways, his desperate need for money, and his involvement in politics, especially in reference to his family, worked great harm to the Church and made him incapable

---

[33] James Hankins, "The Popes and Humanism", in *Rome Reborn: The Vatican Library and Renaissance Culture*, ed. Anthony Grafton (New Haven, Conn.: Yale University Press, 1993), 70.
[34] Storey, *Carnal Commerce*, 70–94.

of responding intelligently to the Protestant Reformation, which began in his reign. As the great Florentine historian and statesman Francesco Guicciardini (1483–1540), who had served the papacy in high-level positions from 1515 to 1527, wrote of it in his famous *History of Italy* (in a passage that was deleted in early editions):

> On these foundations and by these means, raised to secular power, [the popes,] little by little forgetting about the salvation of souls and divine precepts, and turning all their thoughts to worldly greatness, and no longer using their spiritual authority except as an instrument and minister of temporal power, they began to appear rather more like secular princes than popes. Their concern and endeavors began to be no longer the sanctity of life or the propagation of religion, no longer zeal and charity toward their neighbors, but armies and wars against Christians, managing their sacrifices with bloody hands and thoughts; they began to accumulate treasures, to make new laws, to invent new tricks, new cunning devices in order to gather money from every side; for this purpose, to use their spiritual arms without respect; and for this end, to shamelessly sell sacred and profane things. The great wealth spreading amongst them and throughout their court was followed by pomp, luxury, dishonest customs, lust and abominable pleasures: no concern about their successors, no thought of the perpetual majesty of the pontificate, but instead, an ambitious and pestiferous desire to exalt their children, nephews and kindred, not only to immoderate riches but to principalities, to kingdoms; no longer distributing dignities and emoluments among deserving and virtuous men, but almost always either selling them for the highest price or wasting them on persons opportunistically moved by ambition, avarice, or shameful love of pleasure.
>
> And for all these misdeeds, reverence for the papacy has been utterly lost in the hearts of men, and yet their authority is somewhat sustained by the name and majesty, so powerful and effective, of religion; and mightily by the means they have of gratifying great princes and those powerful personages around them, by conferring dignities and other ecclesiastical concessions.[35]

The Reformation began in 1517 as a simple call for an academic disputation on indulgences by an obscure professor, Martin Luther

---

[35] Francesco Guicciardini, *The History of Italy*, trans. Sidney Alexander (New York: Macmillan, 1969), 149.

(1483–1546), at a third-tier German university only founded in 1502. It is somehow appropriate that the trigger for the Reformation should have been the indulgence for the rebuilding of St. Peter's Basilica in Rome, the pope's own church, the symbol of papal power and the great project of the Renaissance papacy in its cultural refashioning. Luther's protests touched deep chords of resentment over abuses, deep desires for reform, and the anticlerical and anti-Roman sentiments that had been percolating for a very long time across Europe. In the hands of a religious genius such as Luther and with its striking invitation to the laity to take charge—and with the emperor Charles V constantly distracted by the threat of the Turks and the French to give his full energies to the matter—it soon took off, spreading rapidly through Germany and beyond. But what must be remembered about the Reformation was that it was not just a protest against abuses, but it was, ultimately, an attack on the Catholic Church in its very nature and doctrines, even at its best. The pope was not merely an incompetent spiritual guide but the antichrist, and Catholicism was itself an evil perversion of true Christianity. The full implications of what the Protestant reformers were trying to achieve was not apparent initially, even to themselves sometimes, but it would in the end create a world very different from the one that preceded it, irretrievably shattering the unity of Europe and creating a whole new culture in the lands where it was triumphant.

The Peace of Augsburg in 1555 ended the religious conflict in Germany for some decades, more or less allowing the territorial rulers to decide their lands' religion (*cuius regio, eius religio*). It left the bulk of Germany in Protestant hands and the Catholic Church in the rest dispirited and in shambles. By Galileo's birth all of Scandinavia, England and Scotland (despite its Catholic queen, Mary Stuart), and half of Switzerland (and all its major towns) had been lost to the Catholic faith, and France and the Low Countries looked as if they would not be long in following them. Even Italy did not seem immune from Protestant influence, and there were a decent number who left Italy for the new faith—Bernardino Ochino (1487–1564), the famous preacher and head of the newly founded Capuchin order, being the most famous. Elia Diodati (1576–1661), a close Geneva-born Parisian friend of Galileo who played an important part in the diffusion of his thought in northern Europe and arranged for the publication of many of Galileo's works, came from such a

family, one that had fled Lucca in northwestern Tuscany for Protestant Geneva.

The Reformation would have a profound effect on Galileo—not
because he ever seemed attracted to its doctrines, but because the
Catholic response to it would end up creating a far more centralized,
defensive, militant, controlling, and coercive Catholic Church than
existed before, and one far less open to change and novelties, including Galileo's. And while one would have thought that the losses of
the Protestant Reformation would have been enough to bring the
Renaissance popes with their obsession with state-building, politics,
and the selfish advancement of their own families to an end, it was
not really until the cataclysmic Sack of Rome that that occurred—at
least to a certain extent and for a while.

The 1527 Sack of Rome was the consequence of the utterly mal-
adroit foreign policy of Pope Clement VII (r. 1523–1534), who allied
himself with the French king Francis I against the Habsburg emperor
Charles V. In May of that year the great city of Rome was captured
and pillaged by a long-unpaid and out of control Imperial army. Its
palaces and churches were plundered, its populace was robbed, tortured, and often murdered, and its women, even nuns, were raped,
as Pope Clement VII and his entourage looked on in dismay and
safety in the Castel Sant' Angelo. Sacrilege was common, with relics and holy objects profaned, and even the tombs of the dead were
desecrated in the pursuit of plunder. Had it not been for the prince
of Orange, who stabled his horses in the Sistine Chapel, the precious
Vatican Library would have been no more, as so many others were
destroyed in the chaos. About ten thousand died during the sack, and
the city was left a shell of itself. The event shocked Europe, but even
many defenders of the papacy saw in it a just judgment of God; and
for some it was a personal turning point.

Only with the rule of Paul III (r. 1534–1549) did the papacy start
to take real leadership in restoring the integrity and purity of the
Church.[36] While Paul III was a notorious promoter of his family's
interests—who had risen to be a cardinal through his sister Giulia,
the mistress of Alexander VI—he also called the reforming Council

---

[36] Elisabeth Gleason, "Who Was the First Counter-Reformation Pope?", *Catholic Historical Review* 81 (1995): 173–84.

of Trent, elevated many reformers to be cardinals, and confirmed the Jesuits and the Capuchins, the two great religious orders of Catholic renewal. Among those reformers who surrounded Paul III were those more conciliatory to the Protestants and who shared some of the same evangelical ideals of the Protestant reformers (often grouped under the name *spirituali*), as well as those more radically opposed and pressing for repression (often grouped under the name *intransigenti*). The failure of the Colloquy of Regensburg (1541) to find points of reunion with some leading Protestant figures opened the way to the creation of the Roman Inquisition in 1542. And the election of the fiery Neapolitan Gian Pietro Carafa (1476–1559), a leading hard-liner, as Pope Paul IV in 1555 spelt the death knell for attempts at reconciliation. Severe and unbending, paranoid in his pursuit of heresy, even within the College of Cardinals (Cardinal Giovanni Morone was imprisoned and put on trial and Cardinal Reginald Pole would have been had he not been in England, where he was protected by Queen Mary Tudor), he engineered a disastrous war against Spain while being blind to his own nephews' crimes.[37] Paul IV's pontificate was a disaster that was saved by Pius IV (r. 1559–1565), who moderated his extreme and counterproductive policies and called back into session the Council of Trent, in abeyance under Paul IV, bringing it to a successful conclusion.

The Council of Trent, meeting in various sessions between 1545 and 1563 (and a long hiatus from 1552 to 1562) in a small town in northern Italy that was yet within "the German lands" (the Holy Roman Empire)—one of the conditions of the Protestants and the emperor Charles V—doctrinally reaffirmed the central tenets of Catholic belief and set out a general program of reform. While the Catholic Church after the Council of Trent is often called "Tridentine" (the adjective for Trent), since its legislation ultimately had such a profound effect upon Catholic life, it should be remembered that although Trent gave a certain doctrinal clarity to Catholicism, it only did so by consensus and compromise between the demands of the princes (who were the greatest advocates for reform) and the papacy (which was very resistant to changes), leaving untouched the more difficult and

---

[37] Miles Patterson, *Pius IV and the Fall of the Carafa: Nepotism and Papal Authority in Counter-Reformation Rome* (Oxford: Oxford University Press, 2013).

divisive points fought over by the various theological schools. (There was a surprising diversity in Catholic theology before the Reformation between different traditions or "schools"—Thomist, Scotist, Nominalist, and so on—not only over specific points but even over fundamental approaches.) The Council of Trent was always a close-run thing, and it was opened only with the greatest difficulty. While the political conflicts between France and the Habsburgs were a problem, it was the papacy's fear of conciliarism, the attempts to limit the power of the papacy by the general councils of bishops, and the loss of papal prerogatives that was the main sticking point. The numbers of bishops attending (or other voting members such as heads of religious orders) were rather low through most of it, and at its peak, at the final session, only 255 were present—and even then the overwhelming majority of them were still Italian. It almost came to an abrupt conclusion many times, and even toward its closing it almost ended in total failure over the topic of the residence of the bishop in his diocese (many bishops had not resided in their dioceses in many years), until Cardinal Morone was sent and arranged a compromise. While reform decrees were promulgated in many areas, two areas were left totally untouched: the reform of the princes, who often had enormous influence over the Church and the right of appointment to major Church offices in their territories, and the reform of the papacy and the Roman Curia.

The papacy and the Roman Curia had been criticized for centuries as the source of many of the worst abuses in the Church. It is quite a jump from the time of Saint Gregory the Great (r. 590–604), "the servant of the servants of God", who wrote to the Byzantine emperor not to consent to his election as pope, to an Innocent III (r. 1198–1216), the Vicar of Christ and the sometime master of kings and emperors, much less to a Boniface VIII (r. 1294–1303), who occasionally wore the Imperial insignia, claiming he was no less emperor as pope, and proclaimed in his bull *Unam sanctam* (1302) the supremacy of the spiritual power over the temporal power. During the Avignon papacy (1309–1378), no secular court was more splendid, and the papal court became a byword for luxury, extravagance, and even decadence—certainly in the eyes of such contemporaries as Saint Bridget of Sweden and Saint Catherine of Siena. The papal bureaucracy more than doubled from three hundred under Pope

Boniface VIII to over six hundred. Also, under the Avignon papacy, the pecuniary advantages of papal primacy and the plenitude of power (*plenitudo potestatis*) began to be realized, and seemingly almost no ecclesiastical benefice—a Church office with an income attached, including everything from that of bishop to that of a lowly clerk—could be safe from papal intervention, manipulation, and taxation, and no Church law seemed secure from a papal dispensation. The papacy was able to draw sums that were close to, and even exceeded, the cash revenues of individual national kingdoms, with Italian bankers and new international methods of credit playing a significant role. While modern popes may rail against the evils of capitalism, papal finance was an essential prerequisite for its early European growth.[38]

In the *Consilium de emendanda ecclesia* (1537), a document put together under the auspices of Pope Paul III by a commission of leading Roman reformers, the source of the evils of the Church was put quite bluntly in the exaltation of papal power and the abuse of that power. "Thus the will of the pope, of whatever kind it may be, is the rule governing his activities and deeds: whence it may be shown without doubt that whatever is pleasing is also permitted."[39] As Vicar of Christ, the pope was above Church law and could arrange things as he thought best. The ease with which dispensations from Church law could be gained in Rome, usually for a fee, made any sort of good order impossible and abuses inevitable, the two greatest abuses being pluralism (holding multiple benefices) and nonresidence in one's benefice. And the greatest abusers in all this were the members of the Roman Curia, which would have been emptied if the traditional norms were enforced. The Council of Trent almost ended in 1562 over the attempt by the reformers to declare the residence of bishops in their dioceses as being of divine law—and therefore not dispensable even by the pope. The compromise arranged by Morone recognized that residence was a divine precept, a more ambiguous term, and the council continued on.[40]

---

[38] Partner, "Papal Financial Policy", 19–20.

[39] John Olin, *The Catholic Reformation: Savonarola to Ignatius* (New York: Loyola Fordham University Press, 1992), 187.

[40] Adam Patrick Robinson, *The Career of Cardinal Giovanni Morone (1509–1580): Between Council and Inquisition* (Burlington, Vt.: Ashgate, 2012), 111–61.

In its reform decrees, the Council of Trent was focused on the renewal of pastoral care, on enhancing the episcopacy and priesthood, and on the obligations of the bishop and parish priest to be models of virtue, holiness, and concern for their flock's spiritual welfare. It was in figures such as Carlo Borromeo (1538–1584), the cardinal-nephew of Pius IV and the first resident archbishop of Milan in eighty years, that the Tridentine model of pastoral care took form; and his personal example of untiring zeal and the decrees from his many synods set the pattern for the rest of the Catholic world to imitate. A zealous bishop faced not merely local vested interests and inertia, the banes of reform generally, but oftentimes Rome herself, who feared episcopal freedom and loss of control. It is ironic that the council which had said very little about the papacy and sidestepped all the old issues and controversies about the extent of papal power, the council which had been much more focused on the bishops and their place in the Church, to enhance the bishop's authority and importance, should become the tool of a revived and newly aggressive papacy. But that is what happened.

While it is true that Trent, rushing to finish up, had assigned to the pope the task of publishing an Index of Forbidden Books, a profession of faith, a catechism, a Breviary (the book which contained the priest's obligatory daily prayers), and a Missal ("the Mass Book"), it had also suggested, without demanding, that if any further definition or declaration were needed in reference to its decrees, the pope should do this in some collegial way, even calling another general council. But when the bull confirming the decrees of Trent was published by Pius IV on June 30, 1564, it gave the pope exclusive jurisdiction to interpret all decrees. Already in December 1563 a Roman congregation had been created to interpret the council and resolve any controversial questions arising from it, one which would later be formalized as the Congregation of the Council in 1588. When the council's canons and decrees were first printed in March 1564, publication of the full proceedings of the council was forbidden for fear it might be used by Protestants for polemical purposes. The pope also forbade the printing of commentaries or notes on the decrees without explicit permission. Later, Pope Paul V would collect all the materials related to Trent into fifty volumes, putting them into the Vatican archives under severe restrictions. All of this severed the council's decisions from their full historical context and meaning, creating the mythical monolith of Trent; and Catholic reform began to take on a more centralized, uniform, and Roman coloration, one which was more generally defensive and apologetic, as well as one increasingly inflexible and static, in which the papacy and Roman Curia displayed a disproportionate significance.[41]

> The language of the Pope is Italian, smacking of the Bolognese patois, which is the worst idiom in Italy.... For the rest, he is a very handsome old man, of middle height, erect, his face full of majesty, a long white beard, more than eighty years old, as healthy and vigorous for his age as anyone can wish, without gout, without colic, without stomach trouble, and not subject to any ailment: of a gentle nature, not very passionate about the affairs of the world; a great builder, and in that respect he will leave in Rome and elsewhere exceptional honor to his memory; a great almoner, I should say beyond all measure. Among other evidences of this, there is not a girl about to marry whom he does not aid in setting up house, if she is of low estate; and in this respect they count his liberality as ready money. Besides that, he has built colleges for the Greeks, and for the English, Scots, French, Germans, and Poles, and has endowed each one with more than ten thousand crowns a year in perpetuity, besides the huge expense of buildings.[42]

Pope Gregory XIII, as described by Montaigne, is a good example of the kind of popes that dominated the second half of the sixteenth century: men zealous and committed to reform, great builders, and full of good works. Pius IV not only brought Trent to a successful conclusion; he created the Roman seminary; published the Tridentine Creed, to which all priests and teachers had to swear; and published a new, less rigorous Index of Forbidden Books. Pius V (r. 1566–1572), later canonized a saint, published the revised Roman Missal, the revised Breviary, and the official Roman Catechism. Gregory XIII (r. 1572–1585) reorganized the permanent papal ambassadors

---

[41] Giuseppe Alberigo, "From Council of Trent to 'Tridentinism' ", in *From Trent to Vatican II: Historical and Theological Investigations*, ed. Raymond F. Bulman and Frederick J. Parrella (New York: Oxford University Press, 2006), 20–30; John W. O'Malley, *Trent: What Happened at the Council* (Cambridge, Mass.: Harvard University Press, 2013), 267.

[42] Montaigne, *Travel Journal*, 75.

into instruments of Church reform and, as we have already seen, labored to improve the clergy by making Rome an intellectual center with his creation or renewal of many colleges and seminaries. In 1586 Sixtus V (r. 1585–1590) set a limit for the number of cardinals (seventy), describing them as assistants and loyal servants of the reformed papacy and its administrative apparatus rather than as semiautonomous Princes of the Church. In 1588 he reorganized the Roman Curia into a modern organization of fifteen congregations, some for spiritual affairs of the Church and some for the temporal affairs of the Papal States, and assigned cardinals to each. Of particular interest was the creation of the Congregation of Rites, which was to oversee worship, requiring all local churches to refer to it on all liturgical questions, including the canonization of saints, thus, over time, creating greater liturgical uniformity under Roman guidance. Sixtus V also formalized the obligation of all bishops to visit the pope at Rome every five years, the visit *ad limina apostolorum* (to the threshold of the apostles), and centralized the process for the nomination of bishops in Rome with the reports of the nuncios rather than with the city or provincial councils governing the nominating process.

Clement VIII (r. 1592–1605), while very devout, was also much weaker in character and far more indulgent to his family than his immediate predecessors. But he did manage to screw up enough courage to absolve Henry of Navarre from the excommunication placed upon him by Pope Sixtus V in 1585 as a relapsed heretic and recognize him as King Henry IV of France, thus bringing an end to the French Wars of Religion (1562–1598), which had devastated the country. This was no mean feat as Spain had her own plans for the French throne and was totally opposed to this reconciliation. She had even threatened to create an independent Spanish Church over it and demanded as a condition of her consent to Clement's election that he oppose it. The powerful Spanish faction in the Roman Curia, and especially among the cardinals, would make it extremely difficult for him to achieve this reconciliation. If Clement VIII failed to get his wishes in so many areas (as we will see later), it was because he had to sacrifice them to achieve this. By this act Clement VIII restored France as a counterweight to Spain and opened it to Catholic reform efforts. A key element in Clement VIII's decision, besides the support given it by Venice and the Medici, was the influence of

Saint Philip Neri (1515–1595), the founder of the Oratorians and the Apostle of Rome, and the other Oratorians who bolstered Clement's weak and indecisive personality. (Clement was especially close to Neri and the Oratorian Church historian Caesar Baronius (1538–1607), who was Clement's confessor.) A Florentine like Galileo who came to Rome in 1533, this "Christian Socrates" exhibited such a singular personality, such a striking originality, kindness, and sense of humor (as well as supernatural gifts), that he charmed not only the cynical Romans but later even the great German writer Goethe (1749–1832), a man not sympathetic to Catholicism who spent pages in his *Travels in Italy* describing the life of this saint who charmed even the animals.

Knowledge and culture he derived, they say, immediately from nature rather than from any stated course of instruction and education: all that other people acquire by patient toil came to him as by inspiration. He further possessed the great gift of discerning spirits, of rightly estimating the qualities and capacities of other men. With the most remarkable penetration his mind also pierced into the events of the world, divining the sequences of things, so that people could not help ascribing to him the spirit of prophecy. He was, moreover, endowed with a mighty power of attraction (*attrattiva*, as the Italians beautifully express it), a power which fascinated not men alone but also animals. The dog of a friend, for example, who came to see him, at once attached itself to Neri, refusing to own its first master or be enticed home with him, though all kinds of allurements were tried, but, held fast by the spell of the eminent man, it would not be separated from him, and after many years ended its life in the bedroom of the master of its choice.[43]

Under Paul V (r. 1605–1621) and Urban VIII (r. 1623–1644), the two most significant popes for Galileo's career, an older and more worldly pattern reasserted itself most strongly, with the papacy again seen as the culmination of family social advancement. While the families of popes had always benefited by their elections and advanced up the social and economic ladder because of it, the Borghese and Barberini took it to a new level. The election of Camillo Borghese

[43]Johann Wolfgang von Goethe, *Goethe's Travels in Italy* (London: George Bell and Sons, 1885), 353.

as Pope Paul V in 1605 was not merely a personal triumph, but the triumph of his family. It had progressed from a very respectable legal family of Siena who had come to practice law in the Roman Curia; and by adept patronage, by a complex and assiduous program of receiving and granting of favors, by the accumulation and buying of offices (which was as common in the other European states as in the Papal States) and the devolution of offices and benefices to clients and kinsmen, by judicious marriages and the cultivation of the right connections, but also by able service to various ecclesiastical and secular leaders, they had become sufficiently well situated to make Camillo's election possible.[44] The Borghese family profited mightily from that election with family members receiving major, and lucrative, secular and ecclesiastical offices and titles, gifts and pensions of every sort, and even outright cash payments from papal revenues, so that the Borghese became great princes in their own right. While the rapacity of the Borghese was notable, it would soon be overshadowed by that of the Barberini of Pope Urban VIII.

While the moral tone of the Curia improved and reform slowly made headway there, it never did so completely; nor does it seem that it made any great change in papal financial policy.[45] Old habits die hard, and sometimes even get resurrected, for the flesh is weak. While simony, the buying and selling of ecclesiastical privileges, disappeared in a technical sense, in substance many of the worst financial abuses continued and even became codified. As the historian Barbara McClung Hallman has shown, the financial reforms sought by the authors of the *Consilium de emendanda ecclesia* of 1537 largely failed and the "cash nexus permeated the hierarchy from top to bottom".[46] Even as ardent a papalist as Robert Cardinal Bellarmine (about whom we will hear a great deal later) was disappointed in the pace of reform. In 1600–1601, at the request of Clement VIII, Bellarmine wrote a memorandum where he criticized the pope for his lack of conformity with the Council of Trent's decrees, particularly criticizing him for leaving

[44] R. Po-chia Hsia, *The World of Catholic Renewal 1540–1770* (Cambridge: Cambridge University Press, 1998), 95–99.

[45] Partner, "Papal Financial Policy", 58.

[46] Barbara McClung Hallman, *Italian Cardinals, Reform, and the Church as Property, 1492–1563* (Berkeley: University of California Press, 1985), 164–68.

dioceses vacant, for his selection of mediocre bishops, for allowing episcopal pluralism and the nonresidence of bishops (including cardinals), and for his leniency in allowing the transfer and resignation of bishops. In 1612, but on his own initiative, Bellarmine sent Pope Paul V a memorandum with four concerns: about bishops who did not reside in their sees, whether to serve the Curia or not; about the excessive frequency of matrimonial dispensations given by the pope; about the accumulation of multiple benefices by single individuals, even with papal permission; and about the luxury and worldliness of bishops and cardinals. Neither document was well received.[47]

> That government is best which is most ordered; and it can be demonstrated that monarchy is more ordered than aristocracy or democracy. All order consists in this, that some should command and others should be subjugated. And order may be discerned not among equals but among those who are superior and inferior. Now, where there is monarchy, there all certainly have some order, since there is no one who is not subject to someone, with the single exception of him who has responsibility for all. For this reason there is the highest order in the Catholic Church, where the people are subjected to the parish priests, the parish priests to the bishops, the bishops to the metropolitans, the metropolitans to the primates, the primates to the pope, the pope to God.[48]

This excerpt from Bellarmine's treatise on the pope's temporal power is a good example of the monarchical view of papal power common among supporters of the papacy, though his more moderate view got him into serious difficulties. In 1582, and again in each of the next three years, the *Summa de potestate ecclesiastica* of Augustinus Triumphus (1243–1328), a promoter of an extreme view of papal power, was reprinted in Rome. And while papal claims to universal monarchy, to have direct power from God in both spiritual and temporal affairs, gained a new lease on life at this time, in point

---

[47]James Broderick, *The Life and Work of Blessed Robert Francis Cardinal Bellarmine, S.J., 1542–1621* (New York: P.J. Kenedy & Sons, 1928), 1:448–56; Stefania Tutino, *Empire of Souls: Robert Bellarmine and the Christian Commonwealth* (Oxford: Oxford University Press, 2010), 263–74, and Alberigo, "From Council of Trent to 'Tridentinism'", 26.

[48]Bouwsma, *Venice and the Defense of Republican Liberty*, 319–20.

of fact, politically, the papacy and the Papal States became a Spanish dependency and would remain so until the election of Galileo's friend, the strongly pro-French Urban VIII, in 1624. Some 25 percent of Rome's population in the early seventeenth century were Spanish or Portuguese, subjects of the Spanish Habsburgs. Papal Rome was too weak to defend itself and too poor to maintain itself, needing Spanish soldiers and ships to protect it, grain from Spanish possessions to keep its population from starvation and riot, and Spanish patronage in benefices, revenues, gifts, and pensions to grease its wheels and burnish its splendor.[49] Spain was the strong right arm of the Church against the Turks and the Protestants, and for this defense of Christendom the papacy actively assisted the Spanish government in the controlling and taxing of its subjects, especially the Church, both in Italy and in Spain.[50] The popes granted to Spain, in particular, three special taxes, the "Three Graces" as they were called: first, there was the *Cruzada*, which was not really a tax but a crusade indulgence first granted for the conquest of Granada in 1492 and which remitted all temporal punishment due to sin and allowed one to eat meat in Lent to any who purchased it; second, there was the *Subsidio*, a tax on the clergy first granted by Pius IV in 1560 to pay for war galleys to fight the Barbary pirates; and third, the *Excusado*, which was another tax on the clergy first granted by Pope Pius V in 1567 to aid Philip II to fight the heretics in Flanders. These taxes, within the pope's prerogative and renewable, became a pillar of the Spanish military budget.[51]

In the Church at large, reform and renewal continued on many levels. While much was inspired by earlier movements and built on traditional Catholic practices which were reinvigorated, it had newer elements as well. In its positive and creative thrust it was a genuine reformation, on par with anything in the Protestant Reformation,

---

[49] Thomas James Dandelet, *Spanish Rome 1500–1700* (New Haven, Conn.: Yale University Press, 2001).

[50] Thomas James Dandelet, "Politics and State System after the Habsburg-Valois Wars", in *Early Modern Italy: 1550–1796*, ed. John Marino (Oxford: Oxford University Press, 2002), 16–19.

[51] David Goodman, *Spanish Naval Power 1589–1665: Reconstruction and Defeat* (Cambridge: Cambridge University Press, 2003), 55–57.

aiming for the improvement of the clergy and pastoral care, the pursuit of holiness and the zeal for souls, the encouragement of social and charitable works in its expanding religious confraternities, and the regeneration of society along Christian lines. This Christianization of society would be a long-term affair, taking centuries of diligent effort and using every persuasive measure available, from the revival of pilgrimage sites and older devotions like the Rosary, to popular missions and new sacramental devotions such as Benediction of the Blessed Sacrament (a short service of adoration of the Holy Eucharist and a blessing with it) and the *Quarant' Ore* or Forty Hours' Devotion (the exposition of the Holy Eucharist for forty hours of continuous worship), to the use of the confessional and parish visitation. In its coercive thrust, it shared with the Protestant Reformation in the social disciplining and the regulation of personal lives, which was part of the fabric of early modern Europe. In its polemical and repressive elements, in its defensive aspect and in its attacks on Protestantism and heresy, in its use of political and military means to achieve its goals, it was a genuine Counter-Reformation, and it included such organs of control as the Roman Inquisition and the Congregation of the Index with its Index of Forbidden Books.

While built on medieval precedents, and separate from the Spanish Inquisition (1478) and the Portuguese Inquisition (1536), both under control of their monarchs, the Roman Inquisition was, unlike its predecessors, a permanent, centralized, and bureaucratic organization situated in Rome and having jurisdiction throughout Italy except for Sicily and Sardinia, which were under the Spanish Inquisition, and the Kingdom of Naples, where it shared authority with the local bishops. (When the Spanish tried to introduce the Spanish Inquisition into Naples in 1547 and in 1563, the people revolted and it did not happen.) The Roman Inquisition, also known as the Holy Office, was preeminent among the fifteen congregations within the Curia, the only one where the pope was its prefect, and its prestige was such that cardinals vied to become members. The committee of cardinals with its assistants, its theological and legal consultants, was also served by local inquisitors and tribunals (more than forty in mainland Italy at the end of the sixteenth century) that worked in association with the local governments throughout the peninsula, whose permission was needed to arrest a suspect.

There was never any great danger that Italy would become Protestant despite the paranoid fears of a Carafa or Michele Ghislieri (later Pius V). Except for the small compact communities of the medieval Waldensians in the Piedmont, who affiliated with the Swiss Protestants in 1532, there were no substantial Protestant churches in Italy, only small conventicles. While there was a clandestine network of preachers, all from religious orders, Ochino being the most famous, who spread the Protestant message by use of coded catch phrases, this was broken up by 1550. While many Italians were attracted to some of the ideas filtering south from the Protestant North (the first Italian translation of a work by Luther was already in 1518), the very complexity of Italian religious life, and the breadth and variegated richness of its reform movements, made it difficult for Protestantism to gain sizable numbers. Only in recent years have we begun to appreciate this complexity, breadth, and richness. In such a world, categories such as *spirituali* and *intransigenti* reveal their profound limitations.[52] Protestantism gained few adherents, and even in cities where it had the greatest presence, such as Venice or Lucca, only 0.2 percent of the population might be fully committed and only 2 percent inclined toward it.[53] The bulk of Italian Protestants neither wished to flee nor to suffer from the Inquisition and lived as "Nicodemists" who conformed externally to the papal Jezebel.

The Inquisition had authority not only over native Italians but even over foreign Protestant scholars, merchants, and visitors staying in Italy, who could be charged with heresy and, if unrepentant, could face death. Had this been strictly enforced, it would have been disastrous for trade and learning, but in most places attempts were made to protect them, though this was less true in the Papal States. The Dutch scholar Henry de Veno (1574?–1613), who would later teach philosophy at the Frisian University of Franeker, born and raised a Protestant (Calvinist), was living quietly in Rome until he was denounced as a

[52] William H. Hudon, "Religion and Society in Early Modern Italy—Old Questions, New Insights", *American Historical Review* 101, no. 3 (June 1996): 783–804; and John Jeffries Martin, "*Renovatio* and Reform in Early Modern Italy" in *Heresy, Culture, and Religion in Early Modern Italy: Contexts and Contestations*, eds. Ronald K. Delph, Michelle M. Fontaine, and John Jeffries Martin (Kirksville, Mo.: Truman State University Press, 2006), 1–17.

[53] Christopher F. Black, *The Italian Inquisition* (New Haven, Conn.: Yale University Press, 2009), 17.

heretic, arrested by the Inquisition, and put on trial in 1597–1598.[54] He was condemned but allowed to abjure his heresies and return to the Catholic fold. After his release, however, he returned home to Holland, to Calvinism and to a successful teaching career. Foreign Protestants had to be careful, and it was not unknown for them to take on other identities to spare themselves from the possibility of arrest by the Inquisition when visiting parts of Italy.

The Inquisition was not just concerned about heresy or religious deviance (and particularly after the 1560s when heretics had become less numerous in Italy) but also with witchcraft, superstition, magic, and immorality of all sorts, particularly in respect to the sacraments as in solicitation in the confessional or in reference to the sacred as in sacrilege and blasphemy. While we would rightly now find abhorrent the pursuit and punishment of deviant beliefs (and the censorship of books for that matter), it was an almost universal practice at the time by both church and state, and by both Catholics and Protestants, with general popular support and very few serious critics. And to give it its due, when compared with the criminal justice system of early modern Europe, the Roman Inquisition holds up rather well, perhaps offering the best criminal justice available in early modern Europe. The rights of the accused were often better guaranteed with procedures clearly outlined in manuals and with an integral right to counsel for the accused, whereas in France and England right to counsel was deliberately excluded. The defendant received a notarized copy of the entire trial, with a reasonable amount of time to reply to the charges, whereas in secular courts the evidence was read against him and he had to defend himself immediately. Torture was rigidly controlled and infrequently inflicted, only in precisely determined cases and only after certain procedures had been followed, whereas judicial torture was routinely used in almost all the major states of Europe in the sixteenth century. Absolution and a merely formal penance was the most common outcome of an Inquisition trial in Italy. Finally, there was a skeptical attitude and leniency toward crimes of witchcraft, sorcery, and spell-casting; and the death penalty was relatively

---

[54] Christoph Lüthy and Leen Spruit, "The Doctrine, Life, and Roman Trial of the Frisian Philosopher Henricus de Veno (1574?–1613)", *Renaissance Quarterly* 56 (2003): 1112–51.

infrequently used, approximately 1250 deaths in a 260-year period. At the very least, the Inquisition spared Italy, Portugal, and Spain the sanguinary witch-hunting craze that so disgraced northern Europe in the sixteenth and seventeenth centuries, killing between forty thousand and sixty thousand, and it, for the most part, only touching the fringes, the Alps and Pyrenees, of the Latin countries.[55]

> The execution was carried out in the Campo de' Fiori, where to terrify him a huge pile of firewood, charcoal, kindling, and more than ten cartloads of pitch had been prepared, and for the occasion a shirt of pitch was made for him that extended from his waist to his feet, black as coal, and then it was put over his naked flesh so that he would not die as quickly, and his life would be consumed in the fire as painfully as possible. He was conducted to the scaffold with a large escort, and made to sit on an iron chair next to the fire, which had already been lit. The usual protest was made on his behalf, as one does for good servants of God, in order to see him repent: that there was still time to obtain grace, but, as soon as he had mounted the iron chair, he threw himself with a great hurry into the burning flames, and buried in them, he died in these earthly flames to spend eternity in those other flames of hell.[56]

This contemporary account from 1595 of the burning in Rome of the Scottish heretic Walter Merse, who had interrupted the Mass at the elevation of the Host, publicly attacking the Holy Eucharist as idolatrous, is an example of where an Inquisition trial could end. The most famous execution for heresy was that of the renegade Dominican philosopher Giordano Bruno (1548–1600), whose death attracted attention all across Europe. After a lifetime travelling across the breadth of the continent (Geneva, Lyon, Toulouse, Paris, London, Oxford, Wittenberg, Prague, Frankfurt, Zurich), often moving in the highest circles, sometimes a Catholic, sometimes a Protestant, this philosopher,

[55] Black, *Italian Inquisition*; John Tedeschi, *The Procesution of Heresy: Collected Studies on the Inquisition in Early Modern Italy* (Binghamton, N.Y.: Medieval and Renaissance Text and Studies, 1991), 8; and Agostino Borromeo, "The Inquisition and Inquisitorial Censorship", in *Catholicism in Early Modern History: A Guide to Research*, ed. John O'Malley (St. Louis: Center for Reformation Studies, 1988), 253–72.

[56] Ingrid Rowland, *Giordano Bruno: Philospher/Heretic* (New York: Farrar, Straus and Giroux, 2008), 11–12.

who was part mystic, part Renaissance magus, and part crackpot and crank, finally met his end on February 17, 1600. After being arrested in Venice and spending eight years on trial and in the Inquisition prison in Rome, he was burned alive, naked and unrepentant, in the Campo de' Fiori (Field of Flowers), the city's market and place of execution. While it is sometimes said that Bruno was condemned for supporting Copernican ideas, his many theological errors, such as denying the Trinity, the divinity of Christ, the Real Presence of Christ in the Holy Eucharist, and so much more, more than explain his horrendous fate. The memory of such a singular and dramatic case lived on, and both Galileo and Bellarmine, who was first a consultant to the Inquisition on the case and then a judge on it after being made a cardinal, were quite aware of it.

> *Item*, that his [Montaigne's] baggage had been inspected by the customs on his entry into the city, and examined right down to the smallest articles, whereas in most of the other cities of Italy these officers were content when you merely offered your baggage for inspection; and that besides this, they had taken from him all the books they had found in order to examine them, which took so long that a man who had anything else to do might well consider them lost; besides, the rules were so extraordinary here that the book of hours of Our Lady, because it was of Paris, not of Rome, was suspect to them, and also the books of certain German doctors of theology against the heretics, because in combatting them they made mention of their errors. In this connection he was grateful for his good luck because, though he had not been warned at all that this was to happen to him, and though he had passed through Germany and was of an inquiring nature, he had no forbidden book in the lot.[57]

While we understand an officious customs inspector looking for illegal drugs or some expensive item hidden to slip through without paying duty, we do not understand and are not accustomed to seeing books as pernicious objects. When Montaigne's books were eventually returned by the Roman authorities only one, a history of the Swiss translated into French was missing, because its translator was a heretic and its preface had been condemned. Also, during his time

---

[57] Montaigne, *Travel Journal*, 73.

in Rome, Montaigne, besides being made an honorary citizen of
Rome and having a very pleasant day looking at the treasures of the
Vatican Library, received back from the Master of the Sacred Palace a
corrected copy of his famous *Essays*. The Master of the Sacred Palace, who licensed books for Rome, was so solicitous and apologetic,
and the criticisms so minor, that in the end no changes were made
in the text. While the censorship of books was also not a new thing
by either church or state, the advent of the Reformation significantly
increased it. By the second half of the sixteenth century it was well-
nigh universal, both among Catholics and Protestants. Thus Rabelais
was banned both by Calvinist Geneva and by the Roman Index.

The early history of the Index of Forbidden Books was a complex one, reflecting the complex situation of the Church herself.[58]
Various Catholic indices of prohibited books had been published in
different places before the first papal Index was promulgated by Pope
Paul IV in 1559. (The Spanish and Portuguese Inquisitions each had
its own Index of Forbidden Books.) Paul's Index was remarkably
strict, included a rather high number of condemnations, and provoked much resistance. It condemned whole authors (e.g., Erasmus)
even when they did not touch upon religion and whole publishers
if they published heretical books. It also banned vernacular Bibles
and the Jewish Talmud. The bishops were totally excluded, and a
confessor could not absolve someone from having possessed a forbidden book, but the penitent had to go to an inquisitor. The Council
of Trent worked on a more moderate version of the Index more in
conformity to humanist ideals, reducing the range of prohibitions
and setting ten rules for censorship. Rule One prohibited all books
previously condemned by popes or general councils. Rule Two prohibited in their entirety the books of the leading heresiarchs (found-
ers of heretical movements) and the writings of heretics that deal
with religion, but allowed for the nonreligious works of those not
heresiarchs to be examined and approved. Rule Three allowed for
translations edited by condemned authors if they contained nothing contrary to correct doctrine. Rule Four allowed the reading of

[58] Gigliola Fragnito, "The Central and Peripheral Organization of Censorship" in *Church Censorship and Culture in Early Modern Italy*, ed. Gigliola Fragnito, trans. Adrian Belton (Cambridge: Cambridge University Press, 2001), 33–35; and Black, *Italian Inquisition*, 158–207.

vernacular Bibles by those who could spiritually profit from them, but only with the written permission of the bishop or inquisitor and only of Bibles prepared by Catholic authors. (This was much freer than before where only the Inquisition in Rome could permit it.) Rule Six prohibited lascivious or obscene books. Rule Seven, while it prohibited the reading of controversial theology in the vernacular without permission, allowed for good vernacular spiritual literature. Rule Eight allowed for a category of books that could just be emended and reprinted, or the expurgation of books and not just their banning. Rule Nine prohibited books dealing with divination and magic ("geomancy, hydromancy, aeromancy, pyromancy, oneiromancy, chiromancy, necromancy, or with sortilege, mixing of poisons, augury, auspices, sorcery, magic arts") and astrology where something was decreed as certain, denying human free will. In Rule Ten it demanded that before a book be printed it needed permission of the Master of the Sacred Palace if printed in Rome, and elsewhere from bishops, inquisitors, or their delegates, with the imprimatur (permission to publish) in the front of the book. It also decreed that there should be inspections of all printers and booksellers, of all those bringing books into a place, or of those inheriting books. Those reading or possessing any heretical book incurred immediate excommunication, and for reading or possessing other kinds of prohibited books, besides incurring mortal sin, they would be severely punished.

The more moderate Tridentine Index came out in 1564 under Pius IV (using the materials from the council), but did not remain long in place before it became more restrictive. When a former inquisitor general, Pius V, became pope, a special Congregation of the Index was set up to regulate the Index in 1571 (later formally and solemnly confirmed by Gregory XIII in 1572), and it was decreed by the Inquisition that all books that had been prohibited in the Index of Paul IV were again prohibited. In 1588 it was among the fifteen congregations created as part of Sixtus V's reorganization of the Roman Curia, and it was given far more impressive cardinals to lead it and the task of identifying books to be prohibited, expurgating those that needed correction, and establishing procedures for the granting of the imprimatur. The preparation of a third index (the so-called Clementine Index) was contentious since it occurred during the crisis over the absolution of Henry of Navarre, and the leading figure in the Inquisition, Cardinal Giulio Antonio Santorio (1532–1602), was not only a leader of the Spanish faction among the cardinals, and so totally opposed to the absolution of Henry of Navarre, but he was also totally opposed to any loosening of censorship. Its publication only occurred in May 1596 after three unsuccessful attempts. It is significant in that it definitively banned vernacular translations of the Bible (as well as selections of it, major adaptions of it, and Bible histories or summaries) despite the desire of the Congregation of the Index and Clement VIII himself to allow for it under the same conditions of Rule Four. Pope Clement VIII's defeat over the use of the vernacular Bible was not his only defeat at this time as we will see, but it shows that even popes could be vanquished within the Curia if they did not develop their own strong factions. And since this Index was to be far more aggressively and strictly enforced, it would have profound consequences, especially for Bible reading in Italy.

Italians were among the first to translate the Latin Bible (Vulgate) into the vernacular, with the printed Malerbi Bible of 1471 as an early example; and the vernacular Bible was widely read in Italy, being one of the most popular spiritual books in the fifteenth and sixteenth centuries.[59] While censorship fit the mentality of the age and was supported by the overwhelming majority of Italians, there was intense resistance to the banning of the vernacular Scriptures. Nonetheless it was vigorously pursued by the authorities, and an enormous number of books were confiscated. And while there was some modification later, allowing for the reading of editions of the Psalms, epistles, and Gospels, with permission of the bishop or inquisitor and only in editions annotated by unquestionably orthodox authors, much had already been burned in the great public bonfires of heretical and lascivious literature that were part of the campaign for spiritual regeneration. Long after the suspension of the prohibition in 1758 by Benedict XIV (the same pope who would remove the prohibition on works advocating heliocentrism), the public mind would associate the vernacular Scriptures with something tainted.

But it was not only heretical books that were affected by the Index, but also books that were seen as obscene or anticlerical, books that

<hr/>

[59] Edoardo Barbieri, "Tradition and Change in the Spiritual Literature of the Cinquecento", in Fragnito, *Church Censorship and Culture*, 125.

attacked Church authority, papal prerogatives, the Roman Curia, or were generally thought as harmful. No book was outside its purview, and no branch of learning. All books needed to be licensed, not just newer ones but reprints of older ones as well; and well-known Italian authors had their works banned, seriously amended, or bowdlerized, including Ariosto, Boccaccio, Castiglione, Dante, Machiavelli, Petrarch, and Aretino. A forbidden book could be seized and burnt publicly, while books that had to be corrected could be seized and perhaps never returned since many books never received their corrections, there being a backlog of such books.

While one could be punished if caught in possession of prohibited books, this pursuit was often hampered by the sheer complexity of it: by its organizational and practical difficulties, by the lack of manpower and the turnover of personnel, by the strength of vested interests, by the inconsistency or ambiguity of some of the directives and the conflicts over jurisdiction, and finally by people's resistance to certain demands that were seen as excessive. Still, it cannot be doubted that the number of books confiscated and destroyed was unprecedented, and in central and northern Italy the pursuit was systematic and ubiquitous.[60] In the South where the local bishops were in charge of orthodoxy and generally were weaker both in reference to local elites and to their own clergy, it was much less severe.

The Tridentine Index also allowed for works to be suspended (banned) *donec corrigatur* (until corrected). Already under Paul IV, in an instruction of February 1559, some allowance was given for the publication of works if they were purged of the names of heretics and all unorthodox material. Besides giving the general criteria of censorship, the Tridentine Index designated the work of correcting to the bishops and inquisitors, and, in some cases, to individuals and universities. Pius V centralized it in the Master of the Sacred Palace (the pope's theologian), and in 1572 the Congregation of the Index was given this task. In 1587, more detailed guidelines were given, though these would not be enforced until 1596 with the promulgation of the Clementine Index. The guidelines were extremely broad, going far beyond protecting against heresy, immorality, and superstition, including anything that might scandalize the faithful

[60] Ibid., 36.

("offensive to pious ears"). It also created a highly unwieldy system of local revisers under the watchful eyes of the local bishop, the local inquisitor, and the Congregation of the Index itself. Attempts were made to divide the suspended works among the various Italian universities or academies (e.g., works of medicine or philosophy to Padua), to particular groups (e.g., humanistic works to the Jesuits), or to learned individuals. While the intention of suspending a work until corrected was to relax the rigor of the Index and allow for more books to be available, from the very beginning it did not work that way, and the vast majority of such books were, for all intents and purposes, permanently banned, the ecclesiastical machinery overwhelmed by the sheer volume of works to be corrected and insufficient structures to do so. One of the most successful of such corrected works would be the 1616 correction of Copernicus' *On the Revolutions of the Heavenly Spheres*.[61]

The pursuit of heresy was not limited to the printed page but was also extended to works of art, as we see in the calling of the Venetian painter Paolo Veronese (1528–1588) before the Inquisition on July 18, 1573. It concerned a massive *Last Supper* he painted in the refectory of the Dominican friary of Saints John and Paul in Venice to replace the Titian *Last Supper* destroyed by fire in 1571:

Q. In this Supper which you painted for San Giovanni e Paolo, what signifies the figure of him whose nose is bleeding?

A. He is a servant who has a nose-bleed from some accident.

Q. What signify those armed men dressed in the fashion of Germany, with halberds in their hands?

A. It is necessary here that I should say a score of words.

Q. Say them.

A. We painters use the same license as poets and madmen, and I represented those halberdiers, the one drinking, the other eating at the foot of the stairs, but both ready to do their duty, because it seemed to me suitable and possible that the master of the house, who as I have been told was rich and magnificent, would have such servants.

[61] Gigliola Fragnito, "The Expurgatory Policy of the Church and the Works of Gasparo Contarini", in *Heresy, Culture, and Religion in Early Modern Italy: Contexts and Contestations*, ed. Ronald K. Delph, Michelle M. Fontaine, and John Jeffries Martin (Kirksville, Mo.: Truman State University Press, 2006): 193–210.

Q. And the one who is dressed as a jester with a parrot on his wrist, why did you put him into the picture?

A. He is there as an ornament, as it is usual to insert such figures.

Q. Who are the persons at the table of Our Lord?

A. The twelve apostles.

Q. What is Saint Peter doing, who is the first?

A. He is carving the lamb in order to pass it to the other part of the table.

Q. What is he doing who comes next?

A. He holds a plate to see what Saint Peter will give him.

Q. Tell us what the third is doing.

A. He is picking his teeth with a fork.

Q. And who are really the persons whom you admit to have been present at this Supper?

A. I believe that there was only Christ and His Apostles; but when I have some space left over in a picture I adorn it with figures of my own invention.

Q. Did some person order you to paint Germans, buffoons, and other similar figures in this picture?

A. No, but I was commissioned to adorn it as I thought proper; now it is very large and can contain many figures.

Q. Should not the ornaments which you were accustomed to paint in pictures be suitable and in direct relation to the subject, or are they left to your fancy, quite without discretion or reason?

A. I paint my pictures with all the considerations which are natural to my intelligence, and according as my intelligence understands them.

Q. Does it seem suitable to you, in the Last Supper of our Lord, to represent buffoons, drunken Germans, dwarfs, and other such absurdities?

A. Certainly not.

Q. Then why have you done it?

A. I did it on the supposition that those people were outside the room in which the Supper was taking place.

Q. Do you not know that in Germany and other countries infested by heresy, it is habitual, by means of pictures full of absurdities, to vilify and turn to ridicule the things of the Holy Catholic Church, in order to teach false doctrine to ignorant people who have no common sense?

A. I agree that it is wrong, but I repeat what I have said, that it is my duty to follow the examples given me by my masters.

Q. Well, what did your masters paint? Things of this kind, perhaps?

A. In Rome, in the Pope's Chapel, Michelangelo has represented
   Our Lord, His Mother, St. John, St. Peter, and the celestial court;
   and he has represented all these personages nude, including the
   Virgin Mary, and in various attitudes not inspired by the most
   profound religious feeling.

Q. Do you not understand that in representing the Last Judgment,
   in which it is a mistake to suppose that clothes are worn, there
   was no reason for painting any? But in these figures what is
   there that is not inspired by the Holy Spirit? There are neither
   buffoons, dogs, weapons, nor other absurdities. Do you think,
   therefore, according to this or that view, that you did well in so
   painting your picture, and will you try to prove that it is a good
   and decent thing?

A. No, my most Illustrious Sirs; I do not pretend to prove it, but I
   had not thought that I was doing wrong; I had never taken so
   many things into consideration. I had been far from imaging such
   a great disorder, all the more as I had placed these buffoons outside
   the room in which Our Lord was sitting.

These things having been said, the judges pronounced that the afore-
said Paolo should be obliged to correct his picture within the space of
three months from the date of the reprimand, according to the judgments and decision of the Sacred Court, and altogether at the expense
of the said Paolo.[62]

The Council of Trent had said very little about art, only speaking
of it as they rushed to finish up in 1563. Defending the usefulness of
images against Protestant iconoclasm, particularly for the illiterate,
it decreed that art should be accurate in its depictions and free of all
superstition, error, or heresy, and not incite its viewers to lust. It had
been thought that Veronese was able to outwit and thumb his nose
at the Inquisition and its demands just by changing the title of the
painting to *The Feast in the House of Levi*, but it now seems reasonably
clear that he did indeed make some significant changes beyond that.[63]

While this kind of censorship could have led to a profound cultural
sterility—and that is the claim of an older historiography that believed

[62] Francis Marion Crawford, *Salve Venetia: Gleanings from Venetian History* (New York: Macmillan, 1905), 2:29–34.

[63] Brian T. D'Argaville, "Inquisition and Metamorphosis: Paolo Veronese and the 'Ultima Cena' of 1573", *RACAR: revue d'art canadienne/Canadian Art Review* 16, no. 1 (1989): 43–48, 99.

that the retrograde heavy hand of the Inquisition and its Spanish ally crushed the genius and individualism of the Italian Renaissance—it does not seem to have done so. While the atmosphere had certainly become more repressive and intolerant, and certain topics were now off-limits, it left the vast majority still untouched. And while the censorship of books did lead to disruptions and delays in publishing, to the disappearance of certain books, and probably scared off not a few authors from writing at all, it does not seem to have cut Italy off from northern learning like an "iron curtain", as some have suggested, or smothered native learning. Italy remained culturally vibrant, and while for the great majority book censorship was relatively effective, for those with the desire and means, a rather extensive clandestine trade in forbidden books developed, and by various subterfuges books of all sorts were available.

> Presiding by divine dispensation over the government of the Church Militant, despite our own unworthiness, and filled with zeal for the salvation of souls which our pastoral office lays upon us, we foster, by token of apostolic favor, certain persons who express their desire for it, and we dispense further graces according as a ripe examination of times and places leads us to judge it useful and beneficial in the Lord.
>
> As a matter of fact, we have lately learned that our beloved sons, Ignatius of Loyola, Pierre Favre, and Diego Laynez, as also Claude Le Jay, Paschase Broet, and Francis Xavier, and further Alphonso Salmeron, and Simon Rodriquez, Jean Codure, and Nicholas Bobadilla, all priests of the cities and dioceses respectively of Pamplona, Geneva, Siguenza, Toledo, Viseu, Embrun, and Placencia, all Masters of Arts, graduates of the University of Paris, and trained for a number of years in theological studies; we have learned, as we say, that these men, inspired, as is piously believed, by the Holy Ghost, have come together from various regions of the globe, and entering into association have renounced the pleasures of this world and have dedicated their lives to the perpetual service of our Lord Jesus Christ, and of ourselves and the other Roman Pontiffs who shall succeed us.[64]

No words (and no twelve names), taken from the beginning of the bull of 1540 *Regimini militantis ecclesiae* of Pope Paul III founding

[64] Olin, *Catholic Reformation*, 203.

the Society of Jesus, have had more significance in the transformation of the Catholic Church of the early modern period than these. While a good part of the Catholic Reform was due to the renewal of older orders, often by new branches, as with the Discalced Carmelites of Teresa of Avila (1515–1582) and John of the Cross (1542–1591), and the Franciscans with the Capuchins, it was really the creation of new orders which were much more focused on the active and missionary element of the faith that the renewal of the Church is due. While other orders such as Saint Vincent de Paul's Congregation of the Mission played a role, the greatest role by far was played by the Jesuits of Ignatius Loyola. While large, the Jesuits were not the largest of the religious orders—it could not match the size of the Capuchins, for instance, who reached thirty thousand members in the seventeenth century—but they were the most adaptable, most mobile, and most educated and intellectual of all the religious orders. One cannot mention the Catholic Reformation without mentioning the Jesuits, who were the soul and engine of it in so many cases.

When Ignatius Loyola (1491–1556), a proud Spanish hidalgo and ex-soldier who became a poor university student at Paris, and his five companions bound themselves on August 15, 1534, the Feast of the Assumption of Mary, at the Chapel of St. Denis in Montmartre, to live a life of service and poverty, to go to Jerusalem, and if that was not possible, to offer themselves to the pope to do whatever work he thought best, they had no idea that they were forming a new religious order. While Ignatius, a practical visionary, had already developed his *Spiritual Exercises*, the manual for the month-long retreat that would lead to a conversion of life and a choice for Christ, he had no fixed ideas for the future. Fortunately for the Church, war between Venice and the Turks did not permit them to go to the Holy Land, and so they turned to the pope, who saw their usefulness. The formal approval as the Society of Jesus in the bull *Regimini militantis ecclesiae* did not envision a large group, for its numbers were restricted to sixty, but that was quickly removed (1543) and the Jesuits grew rapidly: to about one thousand by Ignatius' death in 1556, about thirty-five hundred by 1566, five thousand by 1580, thirteen thousand by 1613, and twenty-two thousand by 1640. Heavily promoted by many of the popes (though not all), free to administer sacraments and preach without the permission of the local bishop or pastor, with a remarkable flexibility, and free of the obligations of the traditional religious orders such as office in choir, with special vows and a unique view of obedience in a highly centralized order, there was no work that the Jesuits did not try their hand at doing and in which they were not supremely successful. They became the great theologians, preachers, philosophers, scientists, spiritual guides, confessors, writers, and teachers of their time.[65] No other group in history has elicited such intense admiration, love, and devotion or such fear, hatred, and envy as have the Jesuits, within the Catholic world as outside it. They also became the great schoolmasters of history, dominating the Catholic educational world. Schools became the primary focus of their activity, and with their thoroughness and system, with their innovation and attentiveness to the students, and charging no fees, they became extremely popular and increased at a rapid rate.

The first Jesuit school was founded at Messina in Sicily in 1548, and they expanded rapidly so that by 1556 they had about 39, by 1580 about 150, by 1607 nearly 300, by 1615 about 370, and more than 800 by the time of their suppression in 1773. The Jesuit schools, while humanist and focused on the classics, also gave an important role to natural philosophy and mathematics. They were the admiration even of Protestant Europe and a significant factor in the reconversion of parts of it to Catholicism. The flagship of their schools was the Roman College. Founded in 1551 as a small school of grammar, humanities, and Christian doctrine, it rapidly grew in size, becoming a major educational establishment, a university really, staffed by their leading scholars. In 1556 it was given the power to confer degrees in philosophy and theology, and with already more than a thousand students it was moved in 1584 to grander quarters paid for by Pope Gregory XIII. It soon grew to more than two thousand students, becoming a major European intellectual center.

While the Catholic Church after the Council of Trent was more uniform and centralized than the Church of the Middle Ages—and while it wanted to promote the image of uniformity and order in opposition to Protestant variety and disorder—it still remained a remarkably variegated reality, far from the monolith it has so often

---

[65] Mordechai Feingold, "Jesuits: Savants", in *Jesuit Science and the Republic of Letters*, ed. Mordechai Feingold (Cambridge, Mass.: M.I.T. Press, 2003), 1–45.

been portrayed as. We see a good example of this when we look at the two religious orders most associated with the Galileo Affair: the Dominicans and the Jesuits.[66]

Both thought of themselves as the intellectual elite of the Church; both considered themselves as the defenders and promoters of the intellectual tradition of Saint Thomas Aquinas. Yet they reacted very differently to the challenges of the times and bore different cultural dispositions. The Dominicans, founded by Saint Dominic (1170–1221) in the thirteenth century in opposition to the rise of heresy and in a time when the revival of cities and the influx of ancient learning was creating new opportunities and dangers for the Church, were the intellectual elite of the Church in the High Middle Ages. They were also major figures in the Inquisition, both in its medieval form and in its more recent manifestations. At the time of the Reformation, however, they seemed to turn more inward and backward, seeing in these troubled times a call to return to an ideal past, a need for ever stricter adherence to their great theologian Saint Thomas Aquinas (1225–1274) and the pagan philosopher Aristotle. They became more defensive toward the world, with their studies ordered to traditional positions and more contemplative ways of life. The Jesuits, ordered totally toward the active life, involved in missions, preaching, writing, and, most importantly, in education, were much more open to the trends of the time. Being at the very frontiers of the Church and the world, of belief and unbelief, they had to develop a certain practicality, flexibility, and conceptual middle ground so as to be true mediators of the faith.

These differences also manifested themselves theologically, particularly during the *De auxiliis* controversy, where from 1588 and the publication of Luis de Molina's *Concordia*, until 1607 when Pope Paul V put a moratorium on any further discussion, the Dominicans and Jesuits were locked in a bitter and very public dispute over grace, predestination, and free will. While the Council of Trent, in opposition to the Protestant understanding of justification by grace alone (*sola gratia*) and their deprecation of human free will, had defined both God's sovereignty and man's freedom and his need to cooperate with

[66] Rivka Feldhay, *Galileo and the Church* (Cambridge: Cambridge University Press, 1995), 73–198.

God's grace, it had left the dynamics of this process vague. In this titanic struggle for "theological hegemony" which was the De auxiliis controversy, the Jesuits put far more emphasis on human freedom while the Dominicans stressed divine omnipotence. The key point of conflict was over Molina's belief in the scientia media, that "middle knowledge" of God by which he infallibly perceives the future contingent acts of man before any absolute divine decree is made. God knows how creatures will react in any circumstances, including under the influence of his grace, if he should decree to concur in them, so an individual's acts are still ultimately his own even under grace. To the Dominicans this dependence on man's future acts was a denial of God's absolute will; and at their most extreme, as under their chief theologian Thomas de Lemos (1555–1629), they denied any reference to God's foreknowledge in the divine decree of predestination but grounded it totally in the divine will.[67]

While not all Jesuits were in complete agreement with Molina, they all opposed the Dominican position, which they believed seemed far too similar to that of the Protestants. This conflict was not a mere petty squabble between religious orders for preeminence or a mere scholastic exercise, the abstract hairsplitting of academics. It had profound practical ramifications, particularly for the Jesuits. As one modern Jesuit theologian has put it: "What was at stake was their whole method of spiritual direction and the whole asceticism of the Spiritual Exercises."[68] Any attack on human freedom and responsibility weakened their whole spiritual and educational program. Debated extensively by both parties when it was brought before the Holy See (1597–1607), it was submitted to the judgment of theological experts six times, and six times the experts condemned the Jesuit position. Yet each time no condemnation occurred. The Jesuits called upon all of their very considerable resources, political and otherwise, to keep themselves from being condemned, as the Dominicans very clearly wanted, and in the end they snatched victory from the jaws of defeat, the participants sent home to await a decision from the pope that never came.

---

[67] Ibid., 183.

[68] Henri Rondet, The Grace of Christ: A Brief History of the Theology of Grace, trans. Tad Guzie (New York: Newman Press, 1967), 327.

The *De auxiliis* controversy did not lead to any condemnation and allowed some intellectual breathing room for the Church, but it did have its consequences, for both religious orders and for the Church at large; and these would also play out in the Galileo case, as we shall see.

Chapter Two

Galileo Surgens: From Obscurity to Celebrity

Galileo Galilei was the first son of Vincenzo Galilei (c. 1525–1591) and Giulia Ammannati (1538–1620). His father had been born in Florence of an ancient and patrician, but now impoverished, Florentine family which had originally borne the name of Buonaiuti. While a Tommaso di Buonaiuti served in the democratic government after the overthrow of the despot Walter of Brienne in 1343, Galileo's most famous ancestor, the first Galileo Galilei, a brother of his great-great-grandfather, was a medical doctor and influential professor at the University of Florence, becoming *gonfaloniere*, or chief magistrate, of Florence in 1445. Vincenzo was an accomplished musician, composer, and noted musical theoretician. His *Dialogue on Ancient and Modern Music* (1581) is perhaps the most influential treatise on music of the late sixteenth century. He was also a member of the Florentine Camerata of Count Giovanni de' Bardi, which attempted to restore music to its classical Greek purity, preferring monody (a single voice) to polyphony (many voices), and which was influential in Florentine cultural circles of the 1570s and 1580s. The work of this circle would be important in the birth of Italian opera. A man of many parts, Vincenzo had a good knowledge of classical languages and Italian literature, was inclined to mathematics and experimentation, had an independent spirit, and a strong and argumentative personality. Nor was he a man impressed by authority. Despite his many talents, trying to raise a large family—for there were two other sons and three or four daughters after Galileo's birth—he was constantly harried by money troubles.

Galileo was educated in his early years at Pisa and later spent some time at the school run by the monks of Vallombrosa, a Benedictine house forty miles southeast of Florence. He even seems to have become a novice there before being taken out by his father over an untended eye infection. As befitting a child of the Renaissance, and as one would expect of his background as a member of the urban patriciate whose social status was set from birth however financially straightened his family may have been, Galileo's early education was classical and humanistic. His biographer, Vincenzo Viviani, remarked on his thorough knowledge of Latin poetry, with Virgil, Ovid, Horace, and Seneca being his favorites; and his knowledge of Greek was sufficiently good for him to begin an Italian verse translation of the mock Greek epic "The Battle of the Frogs and Mice" in 1604 and utilize it for serious study. He knew Petrarch's poems by heart, as well as the satirical poetry of another Florentine, Francesco Berni (1497–1535). He lectured on Dante's *Inferno* at the Florentine Academy at the ripe age of twenty-four and wrote thoughtful commentary on the Italian vernacular poets Ariosto (1474–1533) and Tasso (1544–1595). His tastes in literature and the arts were toward a classical realism, purism, and clarity over a Baroque exaggeration; he desired simplicity, order, measurable forms, clear outlines, and exact proportions as against complexity, imbalance, and confusion; he had an aversion to allegory, mysticism, and rhetorical exaggeration. This aesthetic purity of Galileo may have had as much to do with his rejection of Kepler's laws of planetary motion, with their elliptical orbits over his preferred circular orbits, as did scientific reasons.[1] J. L. Heilbron was right to describe Galileo as a patrician humanist and as a man who—with his many talents, knowledge of literature and music, friendship with the leading families of Florence—could have been many things, but whose life would not have followed the course it did if he had not had to work for a living.[2]

What the poets once sang of the four ages, lead, iron, silver, and gold, our Plato in the *Republic* transferred to the four talents of men, assigning to some talents a certain leaden quality implanted in them, to others iron, to others silver, and to still others gold. If then we are to call any age golden, it is beyond doubt that age which brings forth golden talents in different places. That such is true of this our age he

[1] Leonardo Olschki, "Galileo's Literary Formation", in *Galileo: Man of Science*, ed. Ernan McMullin (New York: Basic Books, 1967), 140–59; and Erwin Panofsky, "Galileo as a Critic of the Arts: Aesthetic Attitude and Scientific Thought", *Isis* 47, no. 1 (March 1956): 3–15.

[2] John L. Heilbron, *Galileo* (Oxford: Oxford University Press, 2010), v–vi.

who wishes to consider the illustrious discoveries of this century will hardly doubt. For this century, like a golden age, has restored to light the liberal arts, which were almost extinct: grammar, rhetoric, painting, sculpture, architecture, music, the ancient singing of the songs to the Orphic lyre, and all this in Florence.[3]

While one must acknowledge an exaggerated exuberance in these 1492 comments of Marsilio Ficino, and a pride in his home city, they do reflect a certain truth about the Italy of the time. In the fifteenth century, Italy seized the intellectual and cultural leadership of Western Europe from its northern neighbors, something that had not been true for centuries, and in many respects it was a "Golden Age". One part was certainly a great artistic revival, in painting, architecture, sculpture, and music, but a very large part was that played by the rise of humanism. Renaissance humanism, in opposition to scholasticism (which came to Italy about the same time as humanism developed there and was in competition with humanism for students and status), was founded on literature, not logic and natural philosophy, and on the *studia humanitatis*, the scholarly study of the great works of classical literature, its grammar, rhetoric, history, poetry, and moral philosophy. The study of this literature, its promoters claimed, had an intrinsic regenerative power and would make men wiser, more virtuous, and more humane, and certainly more eloquent. Nor was it contrary to true Christianity and piety, but even aided them by its humility, its natural virtue, and its attention to the Christian sources (*ad fontes*) of Scripture and the Fathers of the Church, those Greek and Latin Christian thinkers of the first eight centuries who were themselves products of classical learning at its best. While initially focused on ancient Latin literature, it later was broadened to include ancient Greek literature as well. In 1397 the Byzantine scholar Manuel Chrysoloras (c. 1355–1415) came to Florence to teach Greek— with Leonardo Bruni (c. 1370–1444) and Ambrogio Traversari (1386–1439) among his most famous pupils—before teaching at other Italian cities. With the Council of Ferrara-Florence (1438–1445), which briefly reunited the Greek and Latin churches, many other Greek scholars came to Italy, such as Georgius Gemistus Pletho

[3] James Bruce Ross and Mary Martin McLaughlin, eds., *The Portable Renaissance Reader* (New York: Viking Press, 1953), 79.

(c. 1360–1452) and his student Basilios Bessarion (1402–1472), and the knowledge of Greek increased.

In the early fifteenth century professors teaching ancient Greek and Latin literary texts started to appear in Italian universities; and after the middle of the fifteenth century humanism broke out of its purely literary boundaries into all areas of Renaissance culture, including philosophy and science. This was only to be expected considering its great prestige and the fact that almost every scholar received a humanistic training first before going on to university. This humanistic influence can be seen, as the great scholar of Renaissance humanism and philosophy Paul Oskar Kristeller (1905–1999) has pointed out, in the age's emphasis on the dignity of man and his privileged place in the universe, in its tendency to express the concrete uniqueness of one's feelings, opinions, experiences, and surroundings, and in its fundamental classicism, its taste for elegance, neatness, and clarity of style and literary form.[4] Galileo was profoundly formed by this culture, and it should be no surprise that Galileo's two greatest scientific works, *The Two Chief World Systems* and the *Two New Sciences*, should take the form of highly polished literary dialogues, a form favored by classical authors and their imitators. He also reflected his humanist roots by methodically and consistently incorporating into his philosophical arguments literary elements from his favorite poem, Ariosto's *Orlando furioso*, and from similar works.[5]

This city was not long ago blamed for its bad air; but since Duke Cosimo has drained the marshes that are all around, it is good. And it was so bad that when people wanted to confine someone and get him out of the way, they confined him in Pisa, where in a few months it finished him.... Except for the Arno and the beautiful way in which it flows through the town, these churches and vestiges of antiquity, and its private works, Pisa has little distinction and charm. It seems deserted. And in this respect, in the shape of its buildings, its size, and the width of its streets, it is a lot like Pistoia. It is extremely short of water, and the water is bad, for it all has a marshy taste. The men are very poor, and no less haughty, hostile, and discourteous toward

[4] Paul Oskar Kristeller, *Renaissance Thought and Its Sources*, ed. Michael Mooney (New York: Columbia University Press, 1979), 29–31.
[5] Crystal Hall, *Galileo's Reading* (Cambridge: Cambridge University Press, 2014).

foreigners, especially toward the French since the death of a bishop of theirs, Pietro Paolo Bourbon.[6]

While Pisa had a reputation for being uncivil and haughty, and Montaigne in his visit in 1581, as we see above, found little charm in the city itself, he did find a number of kind and courteous people there ("a man who is courteous makes others so"), including some learned professors (medical doctors) attached to the university, though the university itself was not in session. When Galileo enrolled at the University of Pisa in the same year (1581) with the idea of becoming a doctor, a profession as profitable then as now, the city and its university had been much improved. But after three and a half years he left without a degree in 1585. He did, however, find his real vocation, mathematics, during his time there, when he attended the private lectures on Euclid in 1583 by Ostilio Ricci (1540–1603), a friend of his father and tutor to the pages of the Medici court of Grand Duke Francesco then staying at Pisa. Ricci reputedly had been a pupil of the celebrated Niccolò Tartaglia (1499/1500–1577), who did the first Italian translations of the Greek mathematicians Euclid (323–283 B.C.) and Archimedes (c. 287–c. 212 B.C.) and discovered the formula for the solution of cubic equations. Ricci, who later taught at the *Accademia del Disegno*, saw mathematics not as an abstract but as a practical science useful for real-world activities. He passed this practical bent, and his love of Archimedes, on to Galileo. After university Galileo returned to his family in Florence, where for the next four years he taught some and indulged himself in literature and in the intellectual circles of the great city. He developed some new theorems about the centers of gravity of certain solids and invented a hydrostatic balance (*La Bilancetta*) to determine the specific gravities of objects.

In the autumn of 1587, Galileo travelled to Rome seeking to meet those who could advance his career, and particularly to meet with the great Jesuit scientist Christopher Clavius, who for more than fifty years involved himself in the teaching and promotion of the mathematical sciences at the Roman College. Clavius (1538–1612), the

---

[6] Michel Montaigne, *Travel Journal*, trans. Donald M. Frame (San Francisco: North Point Press, 1983), 147–48.

"Euclid of his century", born in Bamberg, Germany, had joined the Jesuits in 1555, being admitted by Ignatius himself. Educated at the ancient University of Coimbra in Portugal and the Roman College, and self-taught in mathematics, he taught that subject at the Roman College from 1563, ceasing his public courses in 1590, after which he only gave specialist courses. He was the architect of the Jesuit mathematics curriculum for all their schools, and initiated and guided an ambitious program of producing commentaries, textbooks, and manuals in all areas of the discipline. He did a commentary of Euclid's *Elements*, and his commentary of the *Sphere* of Sacrobosco (*Sphaera*), the standard introductory textbook on astronomy, would go through sixteen editions in his lifetime (the last in 1611) and was widely used in schools and universities, even non-Jesuit ones, until the middle of the seventeenth century. He was one of the two technical advisors in the commission that reformed the calendar which Pope Gregory XIII promulgated in 1582, and its greatest explicator and defender. His long-term goal was to broaden the range of topics dealt with by mathematics and elevate its status, giving it legitimacy as a true science on par with the rest of the sciences, and his students would form the next generation of Jesuit mathematicians. Under him, by the end of the sixteenth century, the Roman College boasted one of the strongest and most innovative science curricula in Western Europe.[7] The Roman College also housed the elite Academy of Mathematics, a graduate school of sorts which did research and trained the Jesuit order's technical specialists, its mathematics teachers, and its missionaries for areas where scientific specialists were not available. Galileo returned with Clavius' support to help him find a teaching position, leaving with him an original mathematical treatise of his, about which they would correspond in 1588.[8]

In 1588 Galileo gave two lectures on the geography of Dante's *Inferno* at the prestigious Florentine Academy, tutored private students in Florence and Siena, and continued his research. In 1588 he

[7] Paul Grendler, *The Universities of the Italian Renaissance* (Baltimore: Johns Hopkins University Press, 2002), 4.

[8] James M. Lattis, *Between Copernicus and Galileo: Christoph Clavius and the Collapse of Ptolemaic Cosmology* (Chicago: University of Chicago Press, 1994), and Ugo Baldini, "The Academy of Mathematics of the Collegio Romano from 1553 to 1612", in *Jesuit Science and the Republic of Letters*, ed. Mordechai Feingold (Cambridge, Mass.: M.I.T. Press, 2003), 47–98.

failed to gain the chair of mathematics at the University of Bologna, but in 1589 he did win the chair of mathematics at the University of Pisa, with a contract for three years at sixty scudi per year. (In Florence at this time an experienced artisan's annual salary was about fifty scudi.[9]) A key figure in all this was the Marchese Guidobaldo del Monte (1545–1607). As William Shea has pointed out, there was a time when everyone in the Italian scientific community knew who Guidobaldo del Monte was and no one had even heard of Galileo, who was, after all, just an unemployed college dropout at the time.[10] Galileo's meeting with him in 1588 and his continued connection was decisive for his career since not only was Guidobaldo a highly respected mathematician and a well-connected nobleman, but his younger brother Francesco Cardinal Maria del Monte (1549–1627) was also well connected. Francesco Maria (who had Aretino and Titian as godparents) had been educated at the Florentine court with the young princes Francesco and Ferdinando, would serve Ferdinando as an intimate friend and advisor when Ferdinando was a cardinal, and would become Ferdinando's agent at the papal court after he took over the throne. Guidobaldo's support gained Galileo the support of his brother, and since the Grand Duke appointed all teachers at the university, such a connection was invaluable. (Guidobaldo's support would come in handy later on when Galileo applied to the mathematics chair at Padua, where Guidobaldo's cousin, Giovanni Battista del Monte, the Commanding Officer of the Venetian Infantry, lived and where he had much influence.) While the chair of mathematics did not pay well, he was paid better than some, especially given his youth and inexperience, and it was expected that he would take on private pupils to increase his income.[11]

Italy hath many Universities, whereof two are most famous, that of Padoa the cheefe, and of Bologna the next. The University of Bologna is the most auntient, first built (as their recorde sayth) by the Emperour

---

[9] Heilbron, *Galileo*, 391.

[10] William R. Shea, "Guidobaldo del Monte: Galileo's Patron, Mentor and Friend", Edition Open Access, Max Planck Research Library for the History and Development of Knowledge, accessed January 23, 2017, http://www.edition-open-access.de/proceedings/4/5/index.html.

[11] Grendler, *Universities of the Italian Renaissance*, 423.

Theodosious the younger, and long florished under that State (sometymes free, sometymes under priuate Princes) and hath many pre-
vileges from the Popes to whom at last in the tyme of Pope Alexander
the sixth it became subiect. By many Inscriptions in the Princes Pallace
and the publike Schooles, it accknowledgeth Pope Pius the fourth for
a spetiall Benefactor, where also many thinges are written in memo-
rye and honour of the great Jurest Baltlus. The Popes long tyme haue
indowed the same with great stipends for Professors, but espetially for
those of the Imperiall and Papall lawes, and for a cheefe Professor of
Historyes, of whome many learned men have beene upholders of the
Papall power and lawe, against the Emperours.[12]

The Lincolnshire gentleman Fynes Moryson (1566–1630), visiting
Italy in the early 1590s, was right to see Bologna as the oldest Italian
university (and the oldest in Europe), as a center of civil and canon
law, and as being within the Papal States. But it did not go back to
the Theodosius II (401–450), 1088 being the year generally given for
its beginning. Driven by the increased demand for learning and the
need for trained professionals, the number of universities in Europe
went from twenty-nine in 1400 to sixty-three by 1601; and the Italian universities like Bologna, which was the largest Italian university
during the Renaissance in number of students and professors, were
different from their other non-Italian counterparts in important ways.
While Italian universities were similar to those in Spain and northern Europe in that the language of texts, lectures, disputations, and
examinations was Latin and that their professors lectured on the same
books (Aristotle for logic, natural philosophy, and metaphysics; the
works of Hippocrates, Galen, and Avicenna for medicine; the *Corpus juris canonici* and the *Corpus juris civilis* for canon and civil law;
and the *Sentences* of Peter Lombard and the Bible for theology), they
were unlike them in other ways. First, the Italian universities were
generally focused on law and medicine and not on the arts (humanities, logic, and philosophy) and theology. In fact, theology was either
nonexistent or of marginal significance, and when existent generally
taught at local monasteries "off-campus" by members of those religious communities, generally at the *studia generalia* of the mendicant

---

[12] Fynes Moryson, *An Itinerary* (Glasgow: James MacLehose and Sons, 1907), 4:425–26.

orders (primarily Franciscans and Dominicans). Second, the students at Italian universities tended to be older, entering between the ages of sixteen and eighteen and often not finishing until their midtwenties, and they were taught at graduate and professional levels, receiving doctoral degrees but almost never awarded bachelor's degrees. The bachelor's degree had disappeared by about 1400 at Italian universities, whereas the majority of students in the northern universities never got beyond a bachelor's degree. Third, the majority of the professors at Italian universities were married laymen, not clerics, appointed and paid by the civil government, with instruction done at public lectures, open to all, and not part of a residential college as was so often the case at Paris and Oxford. Fourth and finally, the Italian universities were far less organized and cohesive, less a community of scholars and students, and individual professors had greater autonomy, especially if they were well known in their field.[13]

Since philosophy was not geared to theology as in the north, at Italian universities it tended to become much more distinct and autonomous, not a "handmaid of theology". It also became a much more specialized discipline in the later Renaissance, with a greater share of the budget being devoted to it and professors holding chairs exclusively in this and for their whole careers, and not moving on to teach medicine, as was so common earlier.[14] There flourished in Italy a secular Aristotelianism, sometimes called Averroism or Paduan Averroism, after Averroes (1126–1198), the great Arab Muslim commentator of Aristotle, whose naturalism (that nature was more or less an independent system) was a cause of concern to many churchmen. Perhaps the greatest figure in this school was Pietro Pomponazzi (1462–1525), a Mantuan who studied at Padua and then taught there from 1488 to 1496 and 1499 to 1509, when the War of the League of Cambrai closed the university. He later went to the University of Ferrara and then to the University of Bologna in 1512 until his death.

While he lectured in a very traditional, even turgid, scholastic Latin style, which he spiced up with occasional humorous or anticlerical

---

[13] Grendler, *Universities of the Italian Renaissance*, pp. 4–5; Paul Grendler, "The Universities of the Renaissance and Reformation", *Renaissance Quarterly* 57 (2004): 1–42.

[14] Davis L. Lines, "Natural Philosophy in Renaissance Italy: The University of Bologna and the Beginnings of Specialization", *Early Science and Medicine* 6, no. 4 (2001): 267–323.

asides in Italian, Pomponazzi was a bold thinker and a very popular teacher who maintained the affection of many of his students long after their graduation. When he published in 1516 *On the Immortality of the Soul*, he caused enormous controversy for declaring that he did not believe that the immortality of the soul was Aristotle's belief or that it was demonstrable on purely natural grounds, by reason alone—the latter being also the opinion of great scholastic Duns Scotus (c. 1266–1308) and of the leading Thomist of Pomponazzi's day, Cardinal Cajetan (1469–1534). His book was denounced from the pulpit, even publicly burned in Venice by order of the Doge, and a spate of treatises was written against it. During the reign of Pope Leo X, and very likely influenced by Florentine Platonism, the decree *Apostolici regiminis* (1513) of the Fifth Lateran Council (1512–1517) was passed, which condemned as heretics those who denied the immortality of the soul. It also attacked the notion of the double truth—that there could be a contradiction between theology and philosophy—since truth cannot contradict truth. It is generally believed that the decree was directed against the Paduan Aristotelians such as Pomponazzi. More strikingly, the decree also demanded that philosophy professors should not only publicly support the Christian view of the soul but should try to refute arguments philosophically against its immortality ("and the eternity of the world and other things of this sort") insofar as they were able. Pomponazzi, however, would make no attempt to do that, though he was willing to have others do it for him. In his *Apologia* (1518) he restated his position that the soul's immortality could not be rationally demonstrated, since it was contrary to the principles of Aristotelian natural philosophy, though it could and should be accepted as part of divine revelation guaranteed by the authority of the Catholic Church—which he fully accepted. In his *Defensorium* (1519), which included a treatise by the Bolognese Dominican Crisostomo Javelli arguing that immortality of the soul was rationally demonstrable and showing how, he replied that as a professor at the University of Bologna he was only fulfilling the obligations of his position, mandated by the Senate of Bologna and by the pope (in whose territory it lay), to explain Aristotle's views according to natural principles. In the end, he did not retract his opinions nor was he officially condemned; and, even at the height of the controversy in 1518, he not only remained teaching at Bologna, but his contract was renewed, with an increase in salary (double), and he was accorded singular privileges and honors. Also, even much later, long after his death, his treatise on the soul was not put on the Index of Forbidden Books, as did happen in other cases. As one scholar has pointed out, as long as a philosopher treated a contrary philosophical doctrine as only probable or a philosophical theory and stopped short of asserting it as true, he could possibly even argue the apparent philosophical demonstrability of a philosophical position contrary to the teaching of faith.[15] While that interpretation of the decree seems to be correct and helps explain why more was not done against Pomponazzi, having powerful friends and supporters (such as the princely Gonzaga of Mantua and the famous humanist Pietro Bembo, Pope Leo X's Latin secretary) was also not without its significance in this case. This would not be the last time, however, when a seemingly purely philosophical topic would call forth religious criticism and condemnation, as Galileo would find out.[16]

The same kind of Aristotelian naturalism can be seen in Jacopo Zabarella (1533–1589), and later from Cesare Cremonini (1550–1631), who succeeded him. Zabarella, from an ancient and noble family of Padua and educated at the university there, where he taught from 1564 until his death, was free of the scholastic temper of Pomponazzi, and unlike Pomponazzi knew Greek. He was such a master of the Greek texts of Aristotle and their inner principles that the great historian of ideas John Herman Randall (1899–1980) would say of him that it would be unjust to call him a follower of Aristotle: "It is Aristotle himself speaking in the Latin of Padua—not the syllogistic Aristotle with a category for every emergency, but the Aristotle who insists on the primacy of subject matter, of experience. It is the Aristotelian language used to express, not what the ancient Greek thought, but the truth itself."[17] While he was a devout Catholic who

---

[15] Eric A. Constant, "A Reinterpretation of the Fifth Lateran Council Decree *Apostoli regiminis* (1513)", *The Sixteenth Century Journal* 33 (2002): 366.

[16] Paul Oskar Kristeller, *Eight Philosophers of the Italian Renaissance* (London: Chatto & Windus, 1965), 72–90; Jill Kraye, "Pietro Pomponazzi (1462–1525) Secular Aristotelianism in the Renaissance", in *Philosophers of the Renaissance*, ed. Paul Richard Blum (Washington, D.C.: Catholic University of America Press, 2010), 142–60; and Constant, "Reinterpretation of *Apostoli regiminis*".

[17] J. H. Randall, *The School of Padua and the Emergence of Modern Science* (Padova: Editrice Antenore, 1961), 106–7.

spent hours in prayer every day (and was married with children), he too manifested the same kind of independence and naturalism we see in the tradition of Paduan Aristotelianism. While most famous for his work on logic and scientific method, he also wrote extensively on many other topics such as the soul, where he believed, for example, that from the point of natural philosophy the soul could only be seen as mortal. It was for the theologians, he believed, to show its immortality. As far as Galileo, who does mention him a few times in his works, Zabarella is a key part of the well-known Randall Thesis. This was the idea developed by John Herman Randall that Galileo's scientific methodology (and the birth of modern science) is derived from the Paduan Aristotelians and particularly from Zabarella's theory of *regressus*. In this method scientific knowledge is acquired by a twofold process of first ascending from particular effects and experiences to universal causes and then demonstrating, in turn, those effects from their causes. This is also called analysis and synthesis, or resolution and composition, and while it seems to be in Galileo and comes from the Aristotelian tradition (he probably learned it from his study of the works of the Jesuits of the Roman College), what Galileo adds to it, and which is absent from Zabarella and the Aristotelian tradition, is the prime importance given to mathematical reasoning and also to extensive experimentation.

While Aristotle was no longer the overwhelming reality that he had been in the twelfth and thirteenth centuries, he still dominated the philosophical and university worlds in the time of Galileo. More than three thousand editions of his work came out between the invention of printing and 1600 (far more than the five hundred editions of Plato); and there were at least twenty times more commentaries published on his works than on Plato's. The Renaissance humanists did not just revive literary works of the classical age; they revived the philosophical works of antiquity as well. The influx of Greek scholars in the fifteenth century, which had led to a revival of so many Greek works, also led to a revival of the Greek Aristotle. Aldus Manutius published the complete works of Aristotle in Greek between 1495 and 1498 in five folio volumes edited by a team of international scholars including the Englishman Thomas Linacre. In 1497 Niccolò Leonico Tomeo was appointed in Padua to give lectures based on the Greek text of Aristotle. This pattern of privileging the text of Aristotle, particularly the Greek Aristotle, over his commentators, and to interpret Aristotle through Aristotle, spread to other universities. There was also a rediscovery of the Greek commentators on Aristotle, men such as the Aristotelian Alexander of Aphrodisias (fl. A.D. 200) and the Neoplatonists Themistius (fourth century A.D.), Simplicius of Cilicia, and John Philoponus (both sixth century A.D.), as well as a continued interest and reprinting of the medieval commentaries of men such as Saint Thomas Aquinas and Averroes. Aristotelianism's success and continued dominance was not just due to institutional inertia and conservatism, but also to the fact that it had something vital and productive to contribute, to its remarkable adaptability, to its eclecticism, to its capacity to absorb and coexist with other philosophical ideas and trends, and to its capacity to respond to new interests, new insights, and new discoveries. It was extremely diverse—one should rather speak of the plural Aristotelianisms rather than of the singular Aristotelianism—producing such different characters, both in Padua and both laymen, as Jacopo Zabarella, a devout Catholic and friend of the Jesuits who used observation to help determine the truth about nature, and Cesare Cremonini, someone whose Catholic faith was doubted by many and an enemy of the Jesuits, who found truth only in the text of Aristotle itself and supposedly refused to look through Galileo's telescope at a world different from his master's description. And if it had its share of drones and placeholders, it also produced some of the most acute and powerful teachers and thinkers of the day—and no one was free of Aristotelian elements, even Galileo.[18]

In his *Dialogue on the Two Chief World Systems* (1632), Galileo would make fun of Aristotelian philosophers (called Peripatetics because of the *peripatoi*, the "colonnades" or "covered walkways" of the Lyceum, Aristotle's school, where his students met) both in the behavior of the foolish Simplicio, the stock Aristotelian philosopher of the work, and in stories recounted there about various Peripatetics. Stories such as that of the Peripatetic, who when shown by a physician in an anatomical dissection that the origin of the nerves was in the brain and not in the heart as Aristotle stated, replied that

---

[18] Luca Bianchi, "Continuity and Change in the Aristotelian Tradition", in *The Cambridge Companion to Renaissance Philosophy*, ed. James Hankins (Cambridge: Cambridge University Press, 2007), 49–71.

the physician would have clearly and palpably forced him to admit its truth except for the fact that Aristotle's texts so plainly opposed it. Or the story of a professor at a famous university who upon hearing of the telescope, which he had not yet seen, stated that the invention was taken from Aristotle, referring to a passage where Aristotle explains that one can see the stars during the day from a very deep well, with, in his mind, the tube of the telescope taking the place of the well, the thick vapors taking the place of the lenses, and the strengthening of vision being caused by the rays passing through it all. Or the story of the relatively recent "philosopher of great renown", who could opportunistically change his opinion on what Aristotle said on the dangerous topic of the immortality of the soul, and find passages to support it, as was most convenient. As we see in Salviati's comments (who takes the place of Galileo in the *Dialogue*), Galileo certainly believed that Aristotle should be listened to and diligently studied, but not slavishly and ignorantly followed. Knowing as he did that Aristotle judged sensible experience to have priority over human theory (as he has Simplicio state on the First Day and which is repeated a number of times in that section of the *Dialogue*), he also certainly believed that were Aristotle alive he would be opposed to his contemporary defenders:

Do you have any doubt that if Aristotle were to see the new discoveries in the heavens, he would change his mind, revise his books, accept the more sensible doctrines, and cast away from himself those who are so weak-minded as to be very cowardly induced to want to uphold every one of his sayings? Do they not realize that if Aristotle were as they imagine him, he would be an intractable brain, an obstinate mind, a barbarous soul, a tyrannical will, someone who, regarding everybody else as silly sheep, would want his decrees to be preferred over the senses, experience, and nature herself? It is his followers who have given authority to Aristotle, and not he who has usurped or taken it. Since it is easier to hide under someone else's shield than to show oneself openly, they are afraid and do not dare to go away by a single step; rather than putting any changes in the heavens of Aristotle, they insolently deny those which they see in the heavens of nature.[19]

[19] Galileo Galilei, *On the World Systems: New Abridged Translation and Guide*, trans. and ed. Maurice Finocchiaro (Berkeley: University of California Press, 1997), 123.

Ultimately, Aristotle was indispensable, for he was the most systematic and thorough of philosophers, giving the broadest range and the richest content on which to discuss almost any topic. As Sagredo, the third character in the *Dialogue*, put it so eloquently, looking at Simplicio's situation,

> I sympathize with Simplicio, and I see he is much moved by the strength of these very conclusive reasons; on the other hand, he is much confused and frightened by the fact that Aristotle has universally acquired great authority, that so many famous interpreters have labored to explain his meaning, and that other generally useful and needed sciences have based a large part of their reputation on Aristotle's influence. It is as I hear him say: "On whom shall we rely to resolve our controversies if Aristotle is removed from his seat? What other author shall we follow in the schools, academies, and universities? Which philosopher has written on all parts of natural philosophy, so systematically, and without leaving behind even one particular conclusion? Must we then leave the building in which are sheltered so many travelers? Must we destroy that sanctuary ... where so many scholars have taken refuge so comfortably, where one can acquire all knowledge of nature by merely turning a few pages, without being exposed to the adversities of the outdoors? Must we tear down that fortress where we can live safe from all enemy assaults?"[20]

This was not to say that there were not challengers to Aristotle's dominance, a dominance that was, after all, relatively recent in Italy. From the first, the humanists had criticized scholastic Aristotelianism not just for its barbaric Latin style but for it being fundamentally impious, being arid and useless in its argumentativeness, in its obscurity, abstruseness, and abstractions, in its jargon, and in its hairsplitting distinctions and subtleties. There was no reason that elements of both humanism and scholasticsm could not be combined, and there were quite a few who did combine them: a respect for the antique sources, for rhetorical elegance and style, and for practical realities with an appreciation for the content of medieval scholasticism and its method of education. There was always some degree of tension between the two, and in the sixteenth century the polemics between humanism

---

[20] Ibid., 105.

and scholasticism would become far more aggressive and bitter as they moved to northern Europe, particularly as it became involved in the Reformation debates.[21] As Paul Oskar Kristeller has pointed out, while humanism was a movement of great force and had indirect philosophical importance by preparing new concepts and problems—while it contained religious feelings, political views, and moral opinions, and above all a general ideal of education—it was not a philosophical school and contained no metaphysical ideas or speculative systems. At least that was true, he believed, until the later fifteenth century with the Platonic revival around Marsilio Ficino.[22] And, one might also add, that that also was true, to a lesser extent, until the revival of the Hellenistic philosophies of Stoicism, Epicureanism, and skepticism in the sixteenth and seventeenth centuries.[23]

Platonism (including Neoplatonism) had been the dominant philosophy of late antiquity, easily eclipsing Epicureanism and Stoicism, and it played a venerable part in the history of Christian theology. Among early Christian thinkers, East and West, Plato—with his belief in divine providence and immortality of the soul, with his Demiurge who molds the recalcitrant materials of a primitive chaos into a rational, purposeful cosmos, with his focus on the spiritual and transcendent—was far more acceptable than the other philosophers. Later Platonism was even more congenial to a Christian worldview, and Platonic concepts and terminology soon permeated Christian discourse. Certainly, Platonism, or rather the Neoplatonism of Plotinus (c. A.D. 204–270), would have been well known to the Middle Ages through Saint Augustine of Hippo (354–430), who towered over Western theology from the fifth century onward. It would also have been experienced through Pseudo-Dionysius, a late fifth- or early sixth-century Christian Neoplatonist theologian and philosopher, probably Syrian, who was believed to be Saint Paul's convert from the Athenian Areopagus, and whose works had been translated from Greek into Latin in the ninth century. While parts

[21] Erika Rummel, *The Humanist-Scholastic Debate in the Renaissance and Reformation* (Cambridge, Mass.: Harvard University Press, 1995).

[22] Paul Oskar Kristeller, "Florentine Platonism and Its Relation to Humanism and Scholasticism", *Church History* 8 (1939): 201–11.

[23] Jill Kraye, "Philologists and Philosophers", in *The Cambridge Companion to Humanism*, ed. Jill Kraye (Cambridge: Cambridge University Press, 1996), 142–60.

of the Platonic corpus had been translated into Latin, the first two-thirds of Plato's *Timaeus*, for example, it was only with the revival of Greek learning in the fourteenth and fifteenth centuries that we see significant parts of it becoming available. Petrarch praised Plato, calling him the prince of philosophers and comparing him favorably to Aristotle; he even possessed a Greek manuscript of one of his dialogues, though he could not read it, not knowing Greek. It was up to later humanists, educated by the Greek scholars who came West in the late fourteenth and fifteenth centuries, to fill the gap with Galileo's home city of Florence and the Florentine Marsilio Ficino playing a leading role.

Ficino (1433–1499) was the greatest of the Renaissance Platonists, and the chief translator, interpreter, disseminator, and advocate of the Platonic tradition, which he saw not only as the ancient philosophy (*prisca philosophia*) or ancient theology (*prisca theologia*) but also as a *pia philosophia* (a devout philosophy) that could intellectually support and enrich Christianity. Son of the physician to Cosimo de' Medici (1389–1464), who became his lifelong patron, making him tutor to his grandson Lorenzo the Magnificent, Cosimo assisted him in producing the first complete translation of Plato into Latin as well as commentaries on him. (Galileo possessed Ficino's commentary on the *Timaeus*, Plato's major cosmological work.) These translations were widely reprinted and had an enormous impact, being the standard translations for the next three centuries. He also translated the *Enneads* of Plotinus, the next greatest philosopher of this school, and wrote a commentary on it. He even translated lesser figures such as Porphyry (A.D. c. 234–c. 305), Iamblichus (A.D. c. 245–c. 325), and Proclus (A.D. 412–485). Besides his translations he wrote an important synthetic work, *Platonic Theology*, a Platonic *Summa* of sorts, whose primary aim was to defend the immortality of the soul (as the subtitle indicates) but which also discussed a wide range of philosophical topics. This Platonism would influence, for example, the great Church reformer and prior general of the Augustinians (1508–1518) Cardinal Giles of Viterbo (1472–1532), not just in his obtaining the declarations at the Fifth Lateran Council defending the immortality of the soul but also in such works as his commentary on the first seventeen distinctions of the first book of the *Sentences* of Peter Lombard, the standard medieval textbook of theology, *ad mentem Platonis*

(according to the mind of Plato), where he harmonized the philosophical and mythical traditions of antiquity, especially of Platonism,
with the Christian mysteries.[24]

Another area where Ficino had great influence was in the revival
of magic—magic being seen rather as a mastery and manipulation of
nature and other hidden or occult forces than as a kind of hocus-
pocus—which would become a major element in Renaissance culture, though one not much taken seriously by scholars until recently.
Ficino translated the *Hermetic Corpus*, a collection of syncretic and
Platonizing works from the second and third centuries A.D. attributed
to Hermes Trismegistus, a pagan sage who was supposedly a contemporary of Moses, which, while it did not strictly deal with magic, had
a great influence on Renaissance occult and magical circles. More
importantly he published in 1489 his hugely influential *De vita libri tres*
(*Three Books on Life*), which reached almost thirty editions by 1637,
making it the most influential book on magic of its time.[25] Built on
the principle of the interconnectedness of all things, in heaven and
on earth, and a sympathy and correspondence of things between the
two, in the right circumstances one could channel power that God
had implanted in nature for our benefit, even drawing celestial power
into inanimate objects, into talismans and images.

> If you want your body and spirit to receive power from some mem-
> ber of the cosmos, say from the Sun, seek the things which above all
> are most Solar among metals and gems, still more among plants, and
> more yet among animals, especially human beings; for surely things
> which are more similar to you confer more of it. These must both
> be brought to bear externally and, so far as possible, taken internally,
> especially in the day and the hour of the Sun and while the Sun is
> dominant in a theme of the heavens. Solar things are: all those gems
> and flowers which are called heliotrope because they turn towards the
> Sun, likewise gold, orpiment and golden colors,... amber, balsam,
> yellow honey,... the swan, the lion, the scarab beetle, the crocodile,

---

[24] Christopher Celenza, "The Revival of Platonic Philosophy", in Hankins, *Cambridge Companion to Renaissance Philosophy*, 72–96; and James Hankins, "Galileo, Ficino and Renaissance
Platonism", in *Humanism and Early Modern Philosophy*, ed. Jill Kraye and M. W. F. Stone (London: Routledge, 2000), 209–24.

[25] Brian P. Copenhaver, "How to Do Magic, and Why: Philosophical Prescriptions", in
Hankins, *Cambridge Companion of Renaissance Philosophy*, 137.

and people who are blond, curly-haired, prone to baldness, and magnanimous. The above-mentioned things can be adapted partly to foods, partly to ointments and fumigations, partly to usages and habits. You should frequently perceive and think about these things and love them above all; you should also get a lot of light.[26]

The leading Platonist in Galileo's day, Francesco Patrizi (1529–1597), was born of an Italianized Croatian noble family on the Venetian-controlled Dalmatian island of Cherso in the Adriatic. Educated at Venice and Ingolstadt, he entered Padua in 1547, where he would meet Niccolò Sfondrati, later Pope Gregory XIV, and Ippolito Aldobrandini, later Pope Clement VIII. He intended to study medicine but instead turned to the humanities (he would eventually write on poetry, history, and rhetoric), learned Greek, and discovered Plato. The event that changed his life was when he received from a Franciscan friar a copy of Marsilio Ficino's *Platonic Theology*. In Venice in 1571, with a much-expanded version in 1581 at Basel, he published his first major philosophical work, an extensive critique of Aristotle, the *Discussiones peripateticae* (*Peripatetic Discussions*). It included a very critical biography of Aristotle and a discussion of how Aristotle with his physical philosophy was not conformable to that common core of the *prisca sapientia* (ancient wisdom) of such sages as Pythagoras, Hermes Trismegistus, and Plato, which was so congenial to Christianity. Patrizi wished to see the displacement of Aristotle in university teaching, and in 1578 he was called to the University of Ferrara to teach Platonic philosophy, the first-ever chair of Platonic philosophy in a European university. In 1591, dedicating the work to Pope Gregory XIV, and with each of the eleven sections also dedicated to powerful cardinals (all of whom accepted the dedications), he published his *Nova de universis philosophia* (*A New Philosophy of the Universe*), where he elaborated his own grand system that would replace Aristotle. He believed it would not only renew Catholicism but aid in the conversion of heretics and pagans.

Among the dedicatees was Ippolito Cardinal Aldobrandini, who, as Pope Clement VIII, invited him in 1592 to teach Platonic philosophy

---

[26] Marsilio Ficino, *Three Books on Life*, translation with introduction and notes by Carol V. Kaske and John R. Clark (Binghamton, N.Y.: Medieval and Renaissance Texts and Studies, 1989), 247–49.

at the University of Rome in the newly created chair of Platonic philosophy there, where he lectured on the *Timaeus*. While things initially seemed to go well, soon opposition to him developed, with an Inquisition investigation of his major work. Despite Patrizi's best efforts, it was banned in the Index of 1596 until it could be corrected. As we have already seen, the absolution crisis over Henry of Navarre had severely weakened Clement VIII's hand, and to achieve his larger goal of absolving Henry of Navarre he had to sacrifice not only his desire to lessen the strictures of the Index but also the philosopher he himself had called to Rome: Patrizi. The 1590s saw the condemnation in some way of a number of Italian philosophers including Giordano Bruno, Bernadino Telesio, Giovanni Battista della Porta, and Tommaso Campanella. After Patrizi's death Bellarmine recommended that the chair be terminated by Pope Clement VIII, believing that the similarities between Platonism and Christianity made Platonism potentially dangerous to the Catholic faith, but it was filled by Jacopo Mazzoni (1548–1598), Galileo's friend and colleague at Pisa, known for his attempts to reconcile Aristotle and Plato, not to replace Aristotle with Plato.[27]

> I say these public calamities which we suffer are profitable unto us accompanied with an inward fruit and commodity. Do we call them evils? Nay rather they are good if we pluck aside the veil of opinions and cast our eyes to the beginning and end of them; whereof the one is from God, the other for good. The original of these miseries, as I proved plainly yesterday, is of God. That is, not only of the chiefest good, but also of the author, head and fountain of all goodness; from whom it is as impossible than any evil should proceed, as it is for himself to be evil. The divine power is bountiful and healthful, refusing to do or receive harm, whose chief virtue is to do good. Therefore the ancients, though they were void of the knowledge of God, yet having some conceit of him in their brain, called him "Iuppiter a iuvando," that is, of helping; dost thou imagine that he is angry, or choleric, and casts as it were those noisome darts among men? Thou art deceived.

---

[27] Luigi Firpo, "The Flowering and Withering of Speculative Philosophy—Italian Philosophy and the Counter-Reformation: The Condemnation of Francesco Patrizi", in *The Late Italian Renaissance 1525–1630*, ed. Eric Cochrane (London: Macmillan, 1970), 266–84; and F. Furnell, "Jacopo Mazzini and Galileo", *Physis* 14 (1972): 273–94.

Anger, wrath, revenge are names of human affections; and proceeding from a natural frailty and weakness, are incident only to weaklings. But that divine spirit does still persevere in his bounty, and those same bitter pills which he ministers to us as medicines, thou sharp in taste, yet are they wholesome in operation.[28]

This passage from Justus Lipsius' *On Constancy* (1584) helps explain the great popularity of Stoicism in the Renaissance, and particularly from the late sixteenth and early seventeenth century. In a Europe ridden with civil and religious strife and weary of warfare, its focus on perseverance, patience, self-control, and achieving tranquility of soul, despite adversity, was highly attractive, particularly in the first half of the seventeeth century. The ancient Romans had an affinity for Stoicism, even claiming an eminent Roman emperor, Marcus Aurelius, as one of its major authors, so it is not surprising that the revival of antiquity should lead to a revival of interest in it. Petrarch (1303–1374) was enthusiastic for it, as were many other humanists. John Calvin in his humanist youth wrote a commentary on the Roman Stoic Seneca's work on clemency. But its greatest exponent and popularizer was the Flemish humanist Justus Lipsius (1547–1606), whose adaptions of it to Christian tastes and contemporary needs even made it somewhat fashionable. His dialogue (a self-help manual, really) *On Constancy* was extremely popular and was soon translated into Dutch, French, Spanish, German, English, Italian, and Polish. In 1604 he published his *Manuductio ad stoicam philosophiam* (*Guide to Stoic Philosophy*), which gave a detailed account of the origins, history, and doctrines of Stoicism, and *Physiologia* (*Physical Theory of the Stoics*), an in-depth analysis of Stoic physics, which he considered the foundation for its ethics. While the revival of Stoicism had its greatest influence in ethics, it may have also played, in its physics, a part in the cosmological discussions of the sixteenth and seventeenth centuries and in the rise of modern science, with its integration of the earthly and heavenly physics, in its belief that a single animate fluid medium called *pneuma* filled the universe, and

---

[28]Justus Lipsius, *His Second Book of Constancy* (Latin 1584; Englished by John Stradling, 1594; very slightly retouched and annotated by Jan Garrett, 1999–2000), chap. 6, accessed January 23, 2017, http://people.wku.edu/jan.garrett/lipsius2.htm.

in its idea that the planets, being intelligent, directed themselves through this fluid *pneuma*.[29]

> Having named *pleasure* as the *end*, some had used this occasion to slander him, saying that he meant sordid and bodily pleasure. For this reason he makes his own apology, and in order to do away such misrepresentation, he declares even more obviously what sort of pleasure he means and what sort he doesn't. For having thoroughly recommended a sober life, which is satisfied with the simplest and more easily obtainable foods, he goes on to say, "When we say that pleasure is the *end*, we do not mean the sensual or the debauched kind, which terminate in the very moment of enjoyment, and by which the sense are only gratified and pleased, as some ignorant persons who are not of our opinion, or those being enviously bent against us, do thus interpret." We only recognize it as this: "to feel no pain in the body, and to have no trouble in the soul; for it's not the pleasure of continual eating and drinking, nor the pleasure of love, nor that of exotic delicacies, and delicious morsels of large and well-furnished tables, that make a pleasant life; but a sound judgment, assisted by sobriety, and consequently by a serenity and tranquility of mind, which thoroughly inquires into the causes of why we ought to choose or avoid anything; and that drives away all mistaken opinions, or false notions of things, which might raise much perplexity in the soul."[30]

Epicureanism, with its materialism, its denial of the soul's immortality, its hedonism (pleasure was life's goal), its indifference to divinity, and its denial of divine providence seemed least likely of revival, but it also had its attractiveness, as we see in the words of the philosopher Gassendi quoted above. Dante placed Epicurus and his followers in the sixth circle of hell, but much of that may have been due to his unfortunate association with sensual self-indulgence instead of with pleasure in a more refined sense. Its ethics attracted many (Thomas More's Utopians, those models of natural virtue, were ethical Epicureans) and its atomism (that all natural phenomena are explained by a

---

[29] Peter Barker, "Stoic Contributions to Early Modern Science", in *Atoms, Pneuma and Tranquility: Epicurean and Stoic Themes in European Thought*, ed. Margaret J. Osler (Cambridge: Cambridge University Press, 1991), 135–54.

[30] Pierre Gassendi, *On Happiness*, trans. Erik Anderson, accessed August 4, 2014, http://www.epicurism.info/etexts/gassendi_concerninghappiness.html#2.6.

chance collision of atoms in an infinite void space) offered scientists like Galileo a new way of approaching the physical order. Epicurean ideas had been known through the writings of Cicero and Seneca, and the retrieval of the *Lives of Eminent Philosophers* by Diogenes Laertius (third century A.D.), translated into Latin from the Greek by the Camaldolese monk and humanist Ambrogio Traversari (1386–1439), spread in manuscript form and printed in 1472, furthered it. But the most important event in the diffusion of Epicurean ideas was the 1417 finding in a German monastery of a copy of the first-century B.C. Roman poet Lucretius' *De rerum natura* (*On the Nature of Things*) by the humanist and papal secretary Poggio Bracciolini. First printed in 1473, Lucretius' lengthy philosophical poem was the most complete description of Epicureanism from antiquity and was very widely read because of its literary qualities—Galileo possessed two copies of Lucretius in his library—though it took somewhat more time for its philosophical qualities to be accepted.

It took the brilliant French priest-scientist Pierre Gassendi (1592–1655)—whom Edward Gibbon called "the greatest philosopher among literary men, and the greatest literary man among philosophers"—to introduce Epicureanism into the mainstream of European thought and give it respectability by creating a kind of Christian Epicureanism, having stripped it of those elements opposed to the faith. Gassendi has in recent years gained more attention and scholarly interest, and he has generally been remembered for his insightful critique of and dispute with Descartes, but he originally made a name for himself in astronomy. He was the first to observe the transit of Mercury, and there is even a lunar crater named after him. He also corresponded with Galileo, his last letter to him that we possess (October 15, 1637) consoling Galileo for his loss of sight in one eye. (Within a few months Galileo would be totally blind.) Gassendi was a fierce opponent of scholastic Aristotelianism, and his *Exercitationes Paradoxicae adversus Aristoteleos* (1624) revealed his profound disappointment with its emptiness and artificiality. However, the strong opposition he received from the first of seven intended volumes convinced him to abandon the project and move on to the promotion of Epicureanism. The core of Gassendi's natural philosophy was atomism. Atomism went back to the Greeks Leucippus (fl. 450 B.C.) and Democritus (c. 460–c. 370 B.C.), was developed further by Epicurus

and the Epicureans, culminating in Lucretius. For Gassendi all of
the physical phenomena in the universe were explained in terms
of a large but finite number of atoms created and guided by God
along with human free will—denying, therefore, the infinite number
of atoms falling downward in collision with the random "swerves" of
the Epicureans. For him man also possesses a spiritual and immortal
soul unlike the Epicurean material and mortal one. Besides his attacks
on the revival of magic (as in the work against the Englishman Robert Fludd), he also, with Mersenne, another major opponent of magic
and the occult, turned his attentions to the revival of skepticism,
which is our next topic.

> We receive things in one way and another, according to what we are
> and what they seem to us. Now since our seeming is so uncertain and
> controversial, it is no longer a miracle if we are told that we can admit
> that snow appears white to us, but that we cannot be responsible for
> proving that it is so of its essence and in truth; and, with this starting
> point shaken, all the knowledge in the world necessarily goes by the
> board. What of the fact that our senses interfere with each other? A
> painting seems to the eye to be in relief, to the touch it seems flat.
> Shall we say that musk is agreeable or not, which rejoices our sense of
> smell and offends our taste? There are herbs and unguents suitable for
> one part of the body which injure another. Honey is pleasant to the
> taste, unpleasant to the sight.[31]

As Montaigne makes clear from this passage, sense experience can
be a confusing thing, and the difficulty of finding true knowledge in
the world where there is so much false appearance and change has
been with man since he began to philosophize. With philosophical
skepticism, however, it took on a new sophistication. There were
two streams of philosophical skepticism: Academic skepticism, having its origins in Plato's Academy with Arcesilaus (c. 315–241 B.C.)
and Carneades (c. 213–129 B.C.); and Pyrrhonian skepticism, having its origins in Pyrrho of Elis (c. 360–c. 275 B.C.). The Academic
skeptics argued that because of the weakness of the senses and of
reasoning, nothing could be known for certain—that is, that there

---

[31] Michel Montaigne, *The Complete Essays of Montaigne*, trans. Donald M. Frame (Stanford,
Calif.: Stanford University Press, 1948), 452–53.

was no reliable knowledge. While Pyrrhonian skepticism, which received its highest development in the second century A.D. with the Greek physician and philosopher Sextus Empiricus, criticized the Academics as being too dogmatic, it did admit that because of our limitations we should suspend belief. While an awareness of ancient skepticism was not unknown in the Middle Ages (it possessed Augustine's *Against the Academics*, for example), for it to be more widely known it had to await the printing in 1562 of a Latin translation of Sextus Empiricus' *Outlines of Pyrrhonism* and a printing of a Latin translation of his complete works in 1569. But it was the catalyst of the Reformation—which undermined the entire worldview of Christendom and its criterion of truth—for skepticism to become much more common. Luther set the world upside-down—by denying the authority of the pope, the councils, the Church, the Tradition, and even of reason itself, which he called the devil's greatest whore, and placing judgment in the subjective self, in his own conscience instead. As the historian of skepticism Richard Popkin has pointed out, Luther opened a "Pandora's box" that would have the most far-reaching consequences throughout the entire intellectual realm of the West and not just in theology.[32] Philosophical skepticism appealed to those overwhelmed by the complexity of life and the ever-expanding realm of knowledge opening up as new worlds were being discovered and new cultures made known; it appealed to religious folk of all varieties as solidifying the authority of faith and revelation (fideism) against the pretensions of all forms of natural knowledge (philosophical, astrological, and scientific); it appealed to some Catholic apologists who defended themselves and the Church against the appeal to conscience of the Protestant Reformation by focusing on the weakness of reason and the need to follow the well-trodden and safe path of their ancestors; and it appealed to those tired of the often vitriolic religious controversy that marred the age. Pyrrhonism was popularized by Michel de Montaigne (1533–1592)—who suffered through the French Wars of Religion and whose family was religiously divided—in his *Essays*, which was first published in 1580 and very widely read and often reprinted. It was there allied to

---

[32] Richard H. Popkin, *The History of Scepticism: From Savonarola to Bayle* (New York: Oxford University Press, 2003), 5.

a fideism and a conventional Catholicism. In his longest essay, the *Apology for Raymond Sebond*, which contains his famous motto "What do I know?" (*Que sais-je?*), he attacked the presumption of man and defended skepticism as the best response to a world where reason is incapable of finding certainty. This was as true in science as in other areas, as we see in his comments on Copernicus.

> The sky and the stars have been moving for three thousand years; everybody had so believed, until it occurred to Cleanthes of Samos, or (according to Theophrastus) to Nicetas of Syracuse, to maintain that it was the earth that moved, through the oblique circle of the Zodiac, turning about its axis; and in our day Copernicus has grounded this doctrine so well that he uses it very systematically for all astronomical deductions. What are we to get out of that, unless that we should not bother which of the two is so? And who knows whether a third opinion, a thousand years from now, will not overthrow the preceding two?... Thus when some new doctrine is offered to us, we have great occasion to distrust it, and to consider that before it was produced its opposite was in vogue; and, as it was overthrown by this one, there may arise in the future a third invention that will likewise smash the second.[33]

Montaigne's skepticism was founded on his experience of life and the arguments of the ancient skeptical philosophers (not on the intense theological discussions about God's omnipotence, which we will discuss later); and many responded to this spread of skepticism in the early seventeenth century (*la crise pyrrhonienne*) without falling into blind faith or simple conformism. Among these was René Descartes (1596–1650), who sought to find certainty and whose new rationalism would help give birth to modern philosophy. It also included Gassendi and Marin Mersenne (1588–1648), whose pragmatic *via media* or "modified skepticism" allowed imperfect sense knowledge, probable truths about appearances, as sufficient in scientific endeavors.

Mersenne's family were peasant laborers but his intellectual brilliance was noted early on, and he was able to receive a first-class education, attending La Flèche, the Jesuit college which was perhaps the

---

[33] Montaigne, *Essays*, 429.

finest school in Europe, where he was a classmate of Descartes, though he does not seem to have known him there. As an ascetic, devout, and learned Parisian Minim friar who was a major node ("The Mailbox of Europe") in the European-wide correspondence network of scientists, and a facilitator of research and mentor to other scientists, his circle would include such figures as Pierre de Fermat (famous for Fermat's Last Theorem), Etienne, and Blaise Pascal, and Descartes whose scientific career owed a great deal to Mersenne. He would also be very important in spreading Galileo's mathematical ideas in northern Europe in the 1630s. As Mersenne wrote in 1625 in the preface to his *La vérité des sciences* (*The Truth of the Sciences*) there were those, however, who would undercut this noble pursuit of knowledge:

> There is nothing in the world that has such power over our minds as the truth, nor anything that is more contrary to them than falsehood. Indeed the truth has such ascendency over the soul that it forces the mind to give way to all that is veritable, for the understanding has no liberty to reject the truth when it is evident. . . .
>
> Yet, banded against the truth stands a troupe of libertines who, not daring to show their impiety for the fear they have of being punished, try to persuade the ignorant that there is nothing certain in the world because of the continual ebb and flow of everything here below. They try to slip this into the minds of certain young men whom they know to be attracted to libertinism and to all sorts of pleasures and curiosities so that, having discredited the truth in what concerns the sciences and natural things which serve us as ladders to climb toward God, they do the same in what concerns religion. . . .
>
> I believe that there is no Sceptic who, if he takes the time to read this book, will not freely confess that there are many things in the sciences that are veritable and that one must abandon Pyrrhonism if one does not want to lose one's judgment and one's reason.[34]

Mersenne rejected the Aristotelian idea that we can know things in themselves (their essences). While we cannot know everything about something (as God does), still we can know something about something. By coordinating data from several senses at once (which gets

---

us around the deception of the senses), combined with the incontrovertible certainty of common axioms (e.g., "it is impossible for the same thing to be and not be") and the use of syllogisms, we can make probable judgments. Most especially, and the chief theme of *La vérité*, is the certainty of mathematics.[35]

As Amos Funkenstein has pointed out, the combination of elements from Platonism, Stoicism, Epicureanism, and Aristotelianism, and the exclusion of other elements in them, would substantially contribute to the creation of modern science in the seventeenth century.[36] From Platonism would come the idea that the world is best explained through mathematics. From Stoicism would come the view of nature as homogeneous (of the same stuff) and deterministically governed. From Epicureanism would come the idea of the uniformity of efficient natural causes (the forces immediately responsible for bringing something about) and no final cause (the purpose for which a thing exists). And from Aristotle would come an unequivocal description of nature, that the same word has the same meaning when used in a different context.

Galileo mentions Plato many times in his works, favorably, and in a letter to Jacopo Mazzoni on May 30, 1597 (the letter where he first admitted his Copernicanism), he calls Plato their master. Yet, despite his many strong and explicit criticisms of Aristotelianism, sixteen months before his death, in a letter of September 16, 1640, to his friend the Aristotelian philosopher Fortunio Liceti (1577–1657), he stated that in matters of logic he had been an Aristotelian all his life. There are even those who would connect him to the Stoicism of Justus Lipsius through his lectures on the 1604 nova and his association with the Paduan circle of Gian Vincenzo Pinelli.[37] Like many Renaissance thinkers, while Galileo was his own man, he was also influenced by many other sources, and except for skepticism—which was utterly alien to his thought—there were elements of many

---

[35] Peter Dear, *Mersenne and the Learning of the Schools* (Ithaca, N.Y.: Cornell University Press, 1988), 40–41.

[36] Amos Funkenstein, *Theology and the Scientific Imagination: From the Middle Ages to the Seventeenth Century* (Princeton, N.J.: Princeton University Press, 1986), 30–42.

[37] Frances Huermer, "Rubens's Portrait of Galileo in the Cologne Group Portrait", *Notes in the History of Art* 24, no. 1 (Fall 2004): 18–25; and Eileen Reeves, *Art and Science in the Age of Galileo* (Princeton, N.J.: Princeton University Press, 1997).

different philosophies in him, though exactly what and where has been a matter of some controversy and debate. Thus Kristeller would write that Galileo took his distinction between analysis and synthesis in the method of scientific knowledge from Aristotle; his atomism and distinction between primary and secondary characteristics from Democritus (most likely through Epicureanism); his claims for the absolute certainty of mathematical knowledge and that first principles were spontaneously known and produced by the human mind from Plato; and his demand that nature should be understood in quantitative, mathematical terms from the Platonist position of his time.[38] This last point is particularly important since there is much in Galileo which is not Platonic or is even opposed to classical Platonism. Late Renaissance Platonism included a number of things picked up by Galileo (e.g., a view of planetary motion and elemental theory), not least of which was an exalted view of mathematics—certainly far more exalted than among the Renaissance Aristotelians who dismissed the significance of mathematics for physical science—seeing that nature itself is an intrinsically intelligible mathematical structure.[39] The great defender of Galileo's Platonism, Alexandre Koyré, was right to see that for Galileo and his contemporaries this commanding position given mathematics in the study of things natural was the dividing line between Aristotle and Plato, and that the Galilean philosophy of nature could only appear as a triumph of one over the other.[40]

Besides these ancient philosophies, there were also some new philosophies of nature that sought freedom from the strictures of Aristotelianism, such as those of Bernardino Telesio (1509–1588), Giordano Bruno (1548–1600), and Tommaso Campanella (1568–1639), all three of whom were from the Kingdom of Naples. While elements of these anticipated some of the characteristics of later thought, and their attacks on Aristotle weakened his hold on philosophy, they did not produce successors, and all three experienced the harsh hand of authority. Telesio had his main work put on the Index after his death; Bruno lost his life at the stake; and Campanella spent much

[38] Kristeller, *Renaissance Thought*, 64.

[39] Hankins, "Galileo, Ficino and Renaissance Platonism".

[40] Alexandre Koyré, "Galileo and the Scientific Revolution of the Seventeenth Century", *Philosophical Review* 52 (1943): 348.

time in prison and died in exile in France. When Galileo, therefore, began his teaching at Pisa in 1589, the Italian intellectual world was far from stagnant and hidebound, both with the revival of ancient philosophies including Aristotelianism and with the creation of new ones, though one had to be careful about certain topics.

> I had a private talk with a man at Pisa, a good man, but such an Aristo-
> telian that the most sweeping of his dogmas is that the touchstone and
> measure of all solid speculations and of all truth is conformity with the
> teaching of Aristotle; that outside of this there is nothing but chimeras
> and inanity; that Aristotle saw everything and said everything. This
> proposition, having been interpreted a little too broadly and unfairly,
> put him once, and kept him long, in great danger of the Inquisition
> at Rome.[41]

When Montaigne wrote those words about the University of Pisa philosopher Girolamo Borro, the university had already been in exis-tence for nearly 250 years, having been founded in 1343, and was one of the older Italian universities. It was one of sixteen universities that existed in the peninsula in Galileo's day and one of the first authorized to award degrees in theology. While theology was never a strong subject at Italian universities, and even at Pisa never became a dom-inant area of study—at a time when Italian universities were known for their secular tone—Pisa had a much stronger clerical coloration. Closed in 1406 when Florence conquered Pisa, in 1473 Lorenzo the Magnificent moved the University of Florence (except for a handful of positions) to Pisa, where it existed, on and off, until 1526, when it was officially closed again. In 1543 Duke Cosimo I reopened it with three faculties (theology, arts, and law) and twenty professors. By the time Galileo entered it in 1581, it had forty-five professors, with the greatest number in law (fifteen for civil law and seven for canon law), medicine (eight plus one in medical botany), philosophy (six philosophers, two metaphysicians, and two logicians), and the-ology (two), but only one mathematician and one humanist. It was a medium-sized university (in a town of about ten thousand) with an average of six hundred students from 1560 to 1599, awarding an

[41] Montaigne, *Essays*, III.

average of eighty-eight degrees annually from 1570 to 1599, 70 percent of them in law, and with 55 percent of its student body from Tuscany. Many of its students never completed their studies, and, like Galileo, left without degrees. Under the Medici it was strictly forbidden for Tuscans to study outside Tuscany (leaving only Pisa and Siena at which to study), and all graduates and professors had to swear an oath of fealty to the Grand Duke. The Medici had the final say in all appointments. While they attempted to make it into one of the finest universities in Europe, they met with only limited success. It attracted a large number of foreign students, particularly in law, though it never had the same European-wide reputation as Padua or Bologna. In 1566, following the papal bull *In sacrosancta beati Petri* of 1564, the public profession of the Catholic faith became a prerequisite for all professors and for all those receiving a degree, leading to a loss of many German students. (The bull seems often to have been loosely enforced, however, and large numbers of heretic Germans, who were a prime source of income, continued to attend Italian universities.) While Pisa intellectually trailed Bologna and Padua, it may have been equal, or even superior, to other universities its size such as Pavia, Rome, and Ferrara. But it was certainly not the ignorant backwater that some historians have described.[42]

Pisa also seems to have been one of the first universities to offer a study of Platonic philosophy. In 1576 Francèsco Vieri gave extraordinary lectures on Plato along with his ordinary ones on Aristotle. Also filling this task were Galileo's friend Jacopo Mazzoni, who came to Pisa in 1588, and Cosimo Boscaglia (c. 1550–1621), who taught at Pisa from 1600 to 1621, lecturing on the *Timaeus*, *Phaedo*, and *Republic*, and who would reappear at a critical moment in Galileo's career. On the Aristotelian side there had been some distinguished professors at Pisa, and among them were two who likely taught Galileo: Francesco Buonamici (1533–1603) and Girolamo Borro (1512–1592). Both were conservative Aristotelians, though with somewhat different views, and both used experimentation in their theories. While Buonamici was still teaching when Galileo returned, Borro, who had been imprisoned twice by the Inquisition (in 1551 and in 1583) for his

---

[42] Grendler, *Universities of the Italian Renaissance*, 70–77; and Charles Schmitt, "The Faculty of Arts at the Time of Galileo", *Physis* 14 (1972): 243–72.

purely naturalistic and antitheological view of the world, had moved on to Perugia. Neither of them had any great love for the clergy. Once when a friend asked Buonamici if he had read Saint Thomas Aquinas, he replied, "I don't read books by priests."[43] Galileo was also influenced by prominent contemporary teachers from the Jesuit Roman College, the Spaniards Benito Pereira (c. 1535–1610) and Francisco de Toledo (1532–1596), and the German Christopher Clavis, whose notes he used to prepare his own lectures.[44]

Universities did not then primarily function as centers of research, to create new knowledge, but to disseminate old knowledge, the wisdom already achieved from the masters of the past. Certain books were set to be covered by statute. That being said, there seemed to be in practice a certain amount of flexibility in what books and what parts of what books were covered, and even in adding other books alongside the required ones. The best professors did do serious research and writing, and innovative research did go on. The mathematics chair at Pisa included not just mathematics but also astronomy, astrology, and geography. The university statutes state that for the first year they should "read" John of Sacrobosco's *Sphere*, an elementary and introductory text on astronomy from the thirteenth century which was used well into the seventeenth century, Euclid for the second year, and something of Ptolemy for the third. But there was flexibility, and the holder of the chair in the 1570s, Giuseppe Nozzolino, after following the statutory pattern for his first three years, then taught Euclid the next year and Sacrobosco for each of the next four years. The Ptolemy varied greatly, sometimes the *Almagest* (his book on astronomy), sometimes the *Geography*, and sometimes the *Tetrabiblios* (his book on astrology). Galileo taught at Pisa for three years, and according to the archival records he taught in his first year book 1 of Euclid's *Elements*, in his second year book 5 of Euclid, and in his last year book 1 of Euclid and what the records call "*celestium motuum hipotheses*", which is presumably Ptolemy's *Planetary Hypotheses*.[45]

---

[43] Schmitt, "The Faculty of Arts", 263–73. Quotation from Heilbron, *Galileo*, 46.

[44] Alistair Crombie, "Sources of Galileo's Early Natural Philosophy", in *Science, Art and Nature in Medieval and Modern Thought* (London: Hambledon Press, 1996), 149–63; and William Wallace, "Galileo's Jesuit Connections and Their Influence on His Science", in Feingold, *Jesuit Science*.

[45] Schmitt, "Faculty of Arts", 253, 255–63.

Mathematics was very low in the academic pecking order, and distinguished professors in medicine or philosophy earned six or eight times as much. While the sixty scudi per year that Galileo received was not unusual for a new teacher at the university, even after many years of service Galileo could not expect a very high salary.[46] His position there was precarious, and his contract was only for three years, a contract that was unlikely to be renewed since he had not only antagonized other professors and been fined for missing lectures and for not wearing the required academic dress, but he also had offended a powerful member of the ruling Medici family. Galileo, therefore, was in search of a new position and he found one in 1592 at Padua, where the mathematics chair had been vacant since 1588. Despite a rather meager academic background and few achievements, he had powerful supporters who were able to promote his cause. He was appointed by a vote of the Venetian Senate 149 to 8, and having received the necessary permission of the Grand Duke to leave his service at the university, he left for Padua, where he would remain for eighteen years, until 1610.

> But Gentlemen of all Nations come thither in great numbers, by reason of the famous University, which the Emperor Frederick the second, being offended with the City of Bologna, planted here in the yeere 1222, or there abouts, some comming to study the civill Law, other the Mathemetickes &, Musick, others to ride, to practice the Art of Fencing, and the exercises of dancing and activity, under most skilful professors of those Arts, drawn hither by the same reason. And Students have here great, if not too great liberty & priviledges, so as men-slaiers are only punished with banishment which is a great mischiefe, and makes strangers live there in great jealousie of treason to be practised against their lives. The Schoole where the professors of liberall Sciences teach, is seated over against Saint Martins Church, and was of old a publike Inne, having the signe of an Oxe, which name it still retaineth.[47]

So did the great University of Padua appear to Fynes Moryson, who mentions the ancient inn with the sign of an ox which was

---

[46] Ibid., 256.
[47] Moryson, *Itinerary*, 1:156.

extensively renovated into the main lecture hall of the university (Palazzo del Bo), which is why locals called the university *Il Bo* ("the Ox") and its students *bovisti*. Padua, Shakespeare's "fair Padua, nursery of arts" (*Taming of the Shrew*, act 1, scene 1), founded in 1222 (its first sighting in documents), was one of the great universities of Europe, and, many would say, the greatest university of the Renaissance. Tradition has it that it was founded by students and professors from Bologna who sought greater freedom, and its motto became *Universa universis patavina libertas*, "Paduan Freedom is Universal for Everyone." If Bologna was the largest of Italian universities, Padua was the most eminent. By the late fourteenth century Padua was already of European significance, and the conquest of Padua by the Republic of Venice in 1405 did nothing to diminish that. In fact, the Venetian Senate, in taking charge of the university, gave it a significant boost, increasing financial support, endeavoring to secure for it the best teachers, even outsiders, eliminating potential competitors (even in Venice itself), decreeing that all Venetian subjects must study there under pain of a fine of five hundred ducats (a substantial sum), and debarring those who obtained degrees from elsewhere from employment in the Venetian state.

By the close of the fifteenth century its faculty probably had close to sixty professors, divided into thirty-one law professors (seventeen civil law and fourteen canon law) and twenty-eight professors plus one vacancy in the faculty of arts and medicine (twelve professors of medicine, four professors of natural philosophy, three logicians plus one vacancy, two moral philosophers, two metaphysicians, two theologians, two humanists, and a professor of mathematics and astrology) and between nine hundred to one thousand students. During the War of the League of Cambrai, Padua was temporarily lost to Venice and the university suffered greatly. It did not fully reopen until 1517, and with a new faculty it entered its most glorious period, with students coming from all across Europe, Catholic and Protestant, including the American Edmund Davie, who graduated from Harvard in 1674 and received a medical degree from Padua in 1681. While it probably was most famous in the field of medicine— Andreas Vesalius (1514–1564), the father of modern human anatomy, taught surgery and anatomy there from 1537 to 1543—it had leading professors in natural philosophy and law, with a single professor of mathematics, though usually an eminent one. Moryson described both the eminence of mathematics at Padua but also the abundance of private teaching, something Galileo would do a great deal of: "Besydes no place is better than Padoa for the Studye of the Mathematicks, whereof, besides the publike, many priuate teachers may here be founde, and ther want not Students to Consorte and ioyne together, if neede be, to hyre these priuate teachers."[48]

Padua's student body rose from about eleven hundred students in the 1530s to a peak of about sixteen hundred in the 1560s in a town of thirty-five thousand to forty thousand, with a slight decline later that century. An impressive number of famous Renaissance figures studied there, among them Francesco Guicciardini, Pietro Bembo, John Colet, Pico della Mirandola, Tasso, and William Harvey, the English physician who discovered the circulation of blood. Besides the eminence of its teachers, one strength of the university was the protection it gave to its faculty (within reason) and its tolerance of foreign non-Catholic students who, within their lodgings, were free from the queries of the Inquisition and allowed the practice of their religion, including the eating of meat on Fridays and the reading of heretical books. Of these foreign students, the Germans were the most numerous and influential, and more than ten thousand studied at Padua between 1546 and 1630.[49] For a while the university's position was threatened by the Jesuits whose house in Padua opened itself to external students in 1579 with a curriculum in philosophy which overlapped that of the university. In 1582 they added a boarding school for nobles. Many preferred the Jesuit method with its greater supervison over the students and their systematic teaching, and by 1589 it educated some 450 external students and 70 noble boarders as well as the house's Jesuit scholastics. Unfortunately for the Jesuits, in 1589 two of their greatest supporters at the university died, including Jacopo Zabarella, and eventually, they were forced to exclude non-Jesuit students by the Venetian Senate and the threat to the university ceased. The problem was that the Jesuits were seen by the government as *papalini* and *spagnuoli*, supporters of the pope and the Spanish, Venice's enemies. Nor was this perception just held by them as we see in this letter

[48] Ibid., 4:434.
[49] Grendler, *Universities of the Italian Renaissance*, 21–40.

of Alessandro Farnese, the Duke of Parma and head of the Spanish forces in the Low Countries, to his master Philip II: "Your majesty desired me to build a citadel at Maestricht; I thought that a college of the Jesuits would be a fortress more likely to protect the inhabitants against the enemies of the altar and the throne. I have built it."[50]

After Galileo's father's death in 1591, his mother, Giulia, a difficult woman even in the best of times, and his siblings would be a heavy burden on Galileo's time and pocketbook. While he only had to lecture sixty half hours a year, his constant need to make ends meet, and particularly the need to supply dowries for his sisters Virginia and Livia and help out his often impecunious brother Michelangelo, led Galileo to supplement his meager salary of 180 florins with private instruction, taking in boarders and later setting up a small workshop in his house to manufacture instruments. In 1599 his salary was raised to 320 florins and in 1606 to 520 florins. In 1609 it was raised to 1,000 florins, an unheard-of sum for a mathematics professor, and he was given a life appointment, also rather exceptional.

While his first eight years were relatively carefree, from about 1600 he began a family with Marina Gamba, a woman he met in Venice and who bore him three children: Virginia in 1600, Livia in 1601 (both named after his sisters), and Vincenzo in 1606. Galileo never married her and never lived with her. That he did not marry was something not uncommon in Venice, where the cost of dowries was so prohibitive that many upper-class Venetian men and women remained officially single, the women entering convents and the men taking mistresses or frequenting prostitutes, of which there were some ten thousand in Venice as well as some two hundred high-end courtesans for the more discerning and well-to-do. In 1608 the English visitor Thomas Coryat (c. 1577–1617) was greatly impressed by the Venetian noblemen who lived so frugally (soberly dressed in black gowns with no retinue of servants and going to the market themselves to buy their victuals), shocked by the dress of the respectable Venetian women ("with their breastes all naked, and many of them have their backes also naked even almost to the middle"[51]), and fascinated

[50] John Donnelly, "The Jesuit College at Padua: Growth, Suppression, Attempts at Restoration", *Archivum Historicum Societatis Iesu* 50 (1982): 47.

[51] Thomas Coryat, *Coryat's Crudities* (Glasgow: James MacLehose and Sons, 1900), 1:396–97.

with the Venetian courtesans, who were world famous and about which he spoke at length. While he thought Venetians should be daily afraid lest their winking at such licentious uncleanness "should be an occasion to draw down upon them Gods curses and vengeance from heaven, and to consume their city with fire and brimstone,"[52] yet even he was smitten by them.

> For so infinite are the allurements of these amorous Calypsoes, that fame of them hath drawn many to Venice from some of the remotest parts of Christendome, to contemplate their beauties, and enjoy their pleasing dalliances. And indeede such is the variety of the delicious objects they minister to their lovers, that they want nothing tending to delight. For when you come into one of their Palaces (as indeed some few of the principallest of them live in very magnificent and portly buildings fit for the entertainment of a great Prince) you seeme to enter into the Paradise of Venus. For their fairest roomes are most glorious and glittering to behold. The walles round about being adorned with most sumptuous tapistry and gilt leather.... As for her selfe shee comes to thee decked like the Queene and Goddesse of love, in so much that thou wilt thinke she made a late transmigration from Paphos, Cnidos, or Cythera, the auncient habitations of Dame Venus.[53]

Galileo gave his inaugural lecture in December 1592 to great accolades. In his first months in Padua he lived with Gian Vincenzo Pinelli (1535–1601), who had encouraged Galileo to come to Padua and who possessed perhaps the best private library in Italy in the second half of the sixteenth century. According to a 1604 inventory it contained about sixty-five hundred printed books and about eight hundred manuscripts, though it was probably more like nine thousand (or even ten thousand) volumes and about one thousand manuscripts at its height. Born in Naples from a noble Genoese family, amiable, wealthy, and intellectually omnivorous, Pinelli came to Padua at twenty-three to study law and basically never left. Internationally respected as a scholar and a major figure in the Republic of Letters, that network of learned men all across Europe (Catholic and Protestant) to which he was connected by a vast correspondence,

[52] Ibid., 1:399.
[53] Ibid., 1:403–4.

he published nothing himself but labored to assist the researches of others. His home became a center for discussion among the cultivated circles of Venice and Padua, and he put his vast library, and a large collection of scientific instruments of every sort, at the service of savants everywhere—and quite a few visted him from foreign parts. His library was extraordinary, containing texts on a vast range of topics: pagan Greek and Latin literature of every sort, Greek and Latin Fathers of the Church, scholastic theology, popular literature, medicine, botany, poetry, law, political theory, and a good working collection of Greek mathematical and astronomical texts—with works in Latin, Greek, Italian, French, Spanish, Hebrew, and Arabic. Despite Pinelli's orthodox temper and associations with such respectably Counter-Reformation figures as the future Pope Clement VIII, Carlo Borromeo, Bellarmine, and Baronius (to name only the most famous), his library also included books by heretics (generally nondoctrinal ones) and those on the Index of Forbidden Books: having at least ninety prohibited titles and about forty-four banned authors, including books that in many cases had to be purchased abroad and shipped into Italy clandestinely. It aimed to be comprehensive and ecumenical, and under the discerning direction of Pinelli, it achieved that. Unmarried, he divided his time between collecting and studying, and furthering the studies of others. Among those who frequented his home were conservatives such as the renowned Jesuit controversialist and papal diplomat Antonio Possevino (1534–1611) and liberals such as the brilliant polymath and anti-Jesuit Paolo Sarpi. After Pinelli's death most of those from his circle formed a new circle around Antonio Querenghi (1546–1633), a learned Paduan patrician and cleric whose library, while smaller than Pinelli's, was even richer in prohibited books.[54]

The great figure in the university at the time, and the highest paid, was the Aristotelian natural philosopher Cesare Cremonini (1550–1631), who had only arrived at the university a year before from the University of Ferrara, where he had been a friend of the Este rulers

[54] Marcella Grendler, "A Greek Collection at Padua: The Library of Gian Vincenzo Pinelli (1535–1601)", *Renaissance Quarterly* 33 (1980): 386–416; Marcella Grendler, "Book Collecting in Counter-Reformation Italy: The Library of Gian Vincenzo Pinelli (1535–1601)", *Journal of Library History* 16 (1981): 143–51; Heilbron, *Galileo*, 95.

and the poet Tasso. A poet and playwright himself, learned in mathematics and medicine, he was a very popular lecturer, sometimes to more than four hundred students at a time. Intelligent, eloquent, affable, and witty, a defender of the university against the Jesuits who had opened their own school in Padua, threatening the prestige and income of the university, he would become a great friend to Galileo despite their divergent views. In 1608 he even offered his personal financial guarantee on Galileo's behalf when Galileo was seriously overextended. In 1599 Cremonini and Galileo were among the twenty-six founding members of the *Accademia dei Ricovrati*, one of the many scholarly societies common to polite learning in Italy.

Like Sarpi, Cremonini was a perennial object of the Inquisition's attentions and would be endlessly pursued by it.[55] From 1598 on more than eighty Inquisition files were opened on Cremonini, and even after his death, attempts were made to discredit him, particularly in reference to the immortality of the soul. As with Pomponazzi, Cremonini's goal was to reproduce Aristotle's teaching as accurately as possible, and he did not think that Aristotle believed in the immortality of the soul. While refutations of these philosophical opinions existed, he did not believe that it was his job or the job of philosophers to propagate them. That was the job of theologians. In 1613 he published his *Disputatio de coelo* (with the approval of the Paduan inquisitors and the secretary of the Venetian Senate) discussing Aristotle's chief work on cosmology (and ignoring everything else since Aristotle including Galileo). In 1614 Pope Paul V intervened personally through the inquisitor of Padua to have him correct it by providing a refutation of Aristotelian errors. He was threatened that if he did not do so, there would be proceedings made against him on suspicion of heresy and atheism. Cremonini replied that he would not object if the Inquisition had a theologian correct it, but that he could only teach and write what could be demonstrated philosophically, and since he was paid to explain what Aristotle taught, he would have to return his salary if he did anything else. Since he always conformed externally to the demands of the Catholic faith, even observantly so, and had the support of the Venetian authorities, he was able to

55 Edward Muir, *The Culture Wars of the Late Renaissance* (Cambridge, Mass.: Harvard University Press, 2005), 239–43.

maintain his position unharmed. In June 1623, under Pope Gregory XV (1621–1623), the book was put on the Index, and under Pope Urban VIII further attempts were made against him. In June 1625 a call was made to all inquisitors in Italy to seek out any files they might have concerning students of Cremonini and the mortality of the soul.

The papal belief that natural philosophy should be subjected to theology was made even more clear when in 1631 Pope Urban VIII's personal theologian, Agostino Oreggi (whom we will met again in the Galileo case two years later), published a book claiming that Aristotle's true opinion was for the immortality of the human soul. He attributed the idea for his work to a conversation he had with Urban VIII when the pope was still a cardinal. He also claimed that the future pope had said that if Aristotle had really maintained the mortality of the soul, the teaching of his doctrine should be banned in the schools as pernicious and contrary to revealed truth. The final opinion of the theological consultants of the Inquisition on Cremonini's works (July 16, 1632), Urban VIII being the pope, was that his books contained erroneous propositions, even formal heresy, "which do not come from Aristotle's texts, but which he seeks to prove on their own account. Furthermore, in seeking to prove these errors, he provides demonstrations which, according to him, cannot be solved [i.e., refuted]."[56] As Beretta has pointed out, the Inquisition formulated Galileo's crime in practically the same terms the following year.[57]

While many believed that Cremonini denied immortality and that his doctrine promoted atheism among the young, it is difficult to know for certain what Cremonini really thought—and we will probably never know. Certainly, in the texts of his we have, he supports the soul's immortality.[58] One of Cremonini's maxims was supposedly "Intus ut libet, foris ut moris est" (Latin for "In private think as you wish, in public behave as is the custom"). In a world where the literature on dissimulation was quite extensive, and where it was even

---

[56] Francesco Beretta, "Galileo, Urban VIII, and the Prosecution of Natural Philosophers", in *The Church and Galileo*, ed. Ernan McMullin (Notre Dame, Ind.: Notre Dame University Press, 2005), 242–43.

[57] Ibid., 243.

[58] Leonard Kennedy, "Cesare Cremonini and the Immortality of the Human Soul", *Vivarium* 18 (1980): 143–58.

encouraged in the popular books of courtly etiquette, the thoughts of men were not always easy to ascertain. Between the fickleness of princes, the probings of the Inquisition, and the envy of men, dissimulation was a popular option. It was also much safer. As Paolo Sarpi once wrote in a letter, "I am obliged to wear a mask, because no one in Italy may go without one."[59] In any case, there were many who admired and loved Cremonini, and when he died in 1631 of the great plague that was devastating northern Italy, he was buried at the renowned Benedictine monastery of Santa Giustina in Padua, where for many years he had lectured to the monks and to which he left all his possessions.

> I am a Venetian gentleman, and I have never hoped to be known as a literary man. I am well disposed to literary men and I have always protected them. But I do not expect to improve my fortunes or to acquire praises or reputation by becoming famous for my knowledge of philosophy or mathematics, but rather from my integrity and good administration in the magistracies and the governance of the Republic, to which I applied myself in my youth, according to the custom of my ancestors, all of whom wore themselves out and grew old in it. My studies are directed to the knowledge of those things which as a Christian I owe to God, as a citizen to my country, as a noble to my family, as a social being to my friends, and as an honest man and true philosopher to myself.[60]

This letter from Galileo's boon companion Giovanfrancesco Sagredo to the German Catholic banker and supporter of science, Mark Welser (April 4, 1614), sums up well the mind-set of Venetian patriciate, particularly the circle around the Morosini brothers. Galileo frequented the Venetian palace of the Morosini brothers on the Grand Canal, where the intellectual elite of Venice met and which was the "social and intellectual headquarters" of the *giovani*.[61] The Morosini circle also included Sarpi, had briefly included Giordano

---

[59] Jon R. Synder, *Dissimulation and the Culture of Secrecy in Early Modern Europe* (Berkeley: University of California Press, 2009), 86.

[60] William J. Bouwsma, *Venice and the Defense of Republican Liberty* (Berkeley: University of California Press, 1968), 87.

[61] Ibid., 236.

Bruno before his arrest and execution, and included Gianfrancesco Sagredo (1571–1620), whose palace on the Grand Canal Galileo also frequented on his trips to Venice, and with whom he enjoyed the great city in all it opportunities, intellectual and carnal. Sagredo, seven years younger than Galileo, had studied with him in 1597–1598, and while a good scientist, he was also a lover of pleasure, of women, of wine, and of art. Sagredo, whose portrait by Gerolamo Bassano has more recently been identified in the Ashmolean Museum in Oxford, was a Venetian patrician and a loyal supporter of Galileo, assisting him politically and financially, helping him with advice and even in acquiring tools and objects for his work. He remained unmarried and served as Venetian consul in Aleppo in Syria (a major center in the Venetian trading empire) from 1608 to 1611, where his job was to protect, judge, and regulate the Venetian community there, collecting its taxes and duties, controlling its commercial exchanges, keeping in touch with local officials, and informing his Venetian superiors about trade and the political situation. He was immortalized by Galileo as one of three characters in his *Dialogue on the Two Chief World Systems* and in his *Two New Sciences*, where he is described as an intelligent, sharp, and cultured layman, interested and conversant with science.

Also during Galileo's time in Padua his health began to deteriorate. He began to suffer from severe rheumatic attacks in 1604/1605. He suffered from a persistent and severe fever in the summer of 1608 and was bedridden for most of the winter of 1610/1611 and the same summer of 1611. He suffered from gout and heart palpitations, fevers, pains, insomnia, and depressions, and Martin Horky, assistant to Giovanni Antonio Magini (1555–1617), the professor of mathematics at Bologna, thought from a visit in 1610 that Galileo suffered from syphilis (*mal français*). Overindulgence in food and drink, strain, and overwork certainly played a part. The Venetian ambassador to the pope, who had known Galileo before, was surprised in 1615 by his physical deterioration. For the rest of his life Galileo would suffer from reoccurring illnesses, but despite these he would continue his work.[62]

In 1604, a nova, a new star, appeared in the heavens, remaining for eighteen months. In November and December of that year Galileo

[62] Heilbron, *Galileo*, 161–62, 194–95.

gave three well-attended lectures on it showing that it lay beyond the moon, proving that space beyond the moon was not incorruptible and changeless as traditional physics and astronomy believed. Also in 1604 Galileo had his first real brush with the Inquisition when one Silvestro Piagnoni, who had worked for Galileo as a copyist for a year and a half, went before the Inquisition and charged Galileo with a deterministic astrology, casting horoscopes that foretold future events with certainty. He also told the Inquisition that he only saw Galileo go to Mass once, going to the house of "his Venetian prostitute Marina" instead, that he never went to confession, and that he was with Cremonini almost every day.[63] He, however, did admit that he thought that Galileo a believer in matters of the faith if a poor one. While Galileo did cast horoscopes, as we will see in the next chapter, they never claimed to ascertain absolutely the future, and the Inquisition did not pursue the case.

At Padua he taught much the same courses he taught at Pisa. He taught Euclid frequently, the *Sphere* of Sacrobosco, and the *Theorica planetarum*, a book on astronomy by an unknown author of the mid-thirteenth century; and in 1593–1594 and in 1598–1599, he lectured on a Pseudo-Aristotle text from the third century B.C., the *Mechanica* or *Questions of Mechanics*.[64] He complained that the majority of his regular students were medical students who were not really interested in mathematics or able to understand advanced ideas, and he did a great deal of private teaching.[65] He also spent time on other projects such as inventing a machine to pump out water cheaply and devising a very useful geometrical and military compass. He also made his greatest discovery in physics, the law of free fall. But it was another small instrument, the telescope, from which his fame would come and his life be changed beyond all recognition. Galileo had heard about the telescope, invented around 1608 by a Dutchman, from Sarpi, and within a year he had made his own that was much more powerful, one that magnified to the ninth power, which he freely offered to the Venetian state. While an impressed and grateful government granted him life tenure and doubled his salary (though

---

[63] Ibid., 104–5.
[64] Grendler, *Universities of the Italian Renaissance*, 414.
[65] Ibid., 418.

making clear it would be the last increase), what he did with the telescope was more significant: he 'turned it to the stars.

Around the first of December with a telescope now raised to the twentieth power, he began to see that the moon was not the perfect crystalline object of traditional science but mountainous. (Galileo, however, was not the first person to use a telescope to survey the moon. The Englishman Thomas Harriot [1560–1621] did it more than four months before Galileo, and made a drawing of it [August 5, 1609] long before Galileo, but he did not publish it.) Galileo also discovered a slew of stars unknown before. More significantly, he noticed four "planets" (moons) around Jupiter, which he dubbed the Medicean Stars after the Medici ruler and his three brothers. On March 13, 1610, Galileo published his *Sidereus Nuncius* (*The Starry Messenger*) in Venice, and within a week its 550 copies sold out. This short Latin work was dedicated to Cosimo II, his former pupil, the new Grand Duke of Tuscany. It was a sensation.

Now touching the occurrents of the present, I send herewith unto his Majesty the strangest piece of news (as I may justly call it) that he hath ever yet received from any part of the world; which is the annexed book (come abroad this very day) of the Mathematical Professor at Padua, who by the help of an optical instrument (which both enlargeth and approximateth the object) invented first in Flanders, and bettered by himself, hath discovered four new planets rolling about the sphere of Jupiter, besides many other unknown fixed stars; likewise, the true cause of the *Via Lactea*, so long searched; and lastly, that the moon is not spherical, but endued with many prominences, and, which is of all the strangest, illuminated with the solar light by reflection from the body of the earth, as he seemeth to say. So as upon the whole subject he hath first overthrown all former astronomy—for we must have a new sphere to save the appearances—and next all astrology. For the virtue of these new planets must needs vary the judicial part, and why may there not yet be more? These things I have been bold thus to discourse unto your Lordship, whereof here all corners are full. And the author runneth a fortune to be either exceeding famous or exceeding ridiculous. By the next ship your Lordship shall receive from me one of the above-named instruments, as it is bettered by this man.[66]

[66] Logan Pearsall Smith, *The Life and Letters of Sir Henry Wotton*, vol. 1 (1907; repr., Oxford: Clarendon Press, 1966), 486–87.

This letter from March 13, 1610, from the English ambassador to Venice, Sir Henry Wotton, to King James I is a good indication of the effect Galileo's little book had at the time. But the book was also part of Galileo's larger plan to return to Florence in the service of the Medici, for which he had been angling for some time. Beginning in the summer of 1605 and for a few summers he had tutored the Medici heir Prince Cosimo in mathematics. And while he found the atmosphere of the Venetian Republic congenial, serving the Republic had its own burdens. As he wrote to one person, "It is impossible to obtain wages from a republic, however splendid and generous it may be, without having duties attached. For to have anything from the public one must satisfy the public and not any one individual; and so long as I am capable of lecturing and serving, no one in the republic can exempt me from my duty while I receive pay. In brief, I can hope to enjoy these benefits only from an absolute ruler."[67] In a letter of 1610 to Belisario Vinta (1542–1613), the state secretary to the Grand Duke, he wrote that while his income from his official teaching (which was rudimentary) was secure and could be supplemented by private instruction and taking in boarders, all of this impeded his studies and left him little leisure to do research and publish. "Hence if I am to return to my native land, I desire that the primary intention of His Highness shall be to give me leave and leisure to draw my works to a conclusion without being occupied in teaching." Among the books he would complete were two on the system and constitution of the universe, "an immense conception full of philosophy, astronomy, and geometry", three books on local motion, "an entirely new science in which no one else, ancient or modern, has discovered any of the most remarkable laws which I demonstrate to exist in both natural and violent movement", three books on mechanics, and many lesser works on sound and voice, vision and colors, ocean tides, continuous quantities, the motion of animals, and military affairs. Between these books ("dedicated always to my lord") and his inventions ("such inventions as no other prince can match") the prince would not be wasting his money. "But finally, as to the title of my position, I desire that in addition to the title of 'mathematician' His Highness will annex that of 'philosopher'; for I

---

[67] Stillman Drake, *Discoveries and Opinions of Galileo* (New York: Anchor Books, 1957), 65.

may claim to have studied more years in philosophy than months in pure mathematics."[68]

Galileo got his way, and by this he placed himself (and mathematics to a certain extent) not in the inferior realm of unreal abstractions where he had lingered before, the home of the lowly mathematician, but in the more elevated arena of the essences of things where true philosophers dwell. He left Padua in 1610 never to return, leaving behind Marina Gamba, who then got married. His choice to serve his prince in Florence, however, was motivated more than just by his need for leisure to complete his work. It would certainly be a great boon to be free from all that rudimentary and rote teaching, and also to be able to promote what one actually believed and found fascinating. Certainly, the desire to be home also played a significant part. But, however personally satisfying that might be, it was his intellectual ambitions that played a far greater role in his decision. While he was appointed professor of mathematics at the University of Pisa without any obligation to reside or teach there, what was of greater significance, in a world in which connections and status were all important, was that he was also the official "mathematician and philosopher" to a great reigning prince, and a prince known for his Catholic connections. It gave his work a certain prestige and acceptance that even being a professor at the great University of Padua could not. Also, while the Venetian Republic was an agreeable atmosphere in which to work, its history of opposition to the papacy, as seen, for example, in the Venetian Interdict crisis of 1606–1607, to a certain extent cut him off from the Catholic mainstream. While he was at Padua, Galileo had become friendly with a number of Jesuits including the able mathematician Giuseppe Biancani, of whom we will hear more later. (A favorite student of Galileo's, John Schreck, would later join the Jesuits, becoming a missionary in China.)[69] If Galileo was trying to advance his career and his ideas, Padua was not the place—and being friends with Sarpi and Cremonini would not help. Padua was invariably associated with Aristotelian naturalism, and that was no asset for Galileo. Leaving Padua was more than just a

[68] Ibid., 60–65.

[69] Pasquale D'Elia, "The Spread of Galileo's Discoveries in the Far East (1610–1640)", *East and West* 1, no. 3 (October 1950): 156–63.

mundane career move; it was a political and intellectual choice. And as we leave Padua, it is most appropriate that we end with that great paean of the ancient university from John Herman Randall:

> What Paris had been in the thirteenth century, and Oxford and Paris together in the fourteenth, Padua became in the fifteenth: the center in which ideas from all Europe were combined into an organized and cumulative body of knowledge.... Scientific Padua felt the effects of the same Humanistic impulse and the same revival of learning that were inspiring Florence in the second half of the quattrocento. To the challenge of Ficino's Platonism it responded by proving that Aristotle as well as Plato spoke Greek. To Ficino's attack on their traditional Averroism, with its fatalism and its strange conception of human nature that minimized all that was personal and individual, the Paduans replied, not by accepting his Platonic religious modernism, but by reorganizing their own naturalistic and scientific thought around a more individualistic conception of man and his destiny. To the Florentine Platonic Humanism they opposed an Aristotelian Humanism close to the naturalism of Aristotle himself and fitting well with their dominant scientific interests. Not until Spinoza and the eighteenth-century Newtonians does there appear another figure who manages to effect so "modern" a blend between Humanism and scientific naturalism as Pomponazzi and Zabarella.[70]

Not long after he returned to Florence, Galileo discovered that Venus went through phases like the moon, but that these phases could only be explained if Venus was going around the sun and not the earth. This strengthened the case for the Copernican worldview. Because early telescopes were limited, many people had doubts about the validity of Galileo's findings. Galileo distributed his own superior telescopes, and directions to use them, to leading figures so as to confirm his discoveries. But even then, it could be a tricky thing. On a visit to Bologna after his return to Florence, he failed to be able to show Magini and the other learned guests the Medicean Stars. The Imperial mathematician Johannes Kepler (1571–1630) was the first significant astronomer to endorse Galileo. He wrote defending the use of the telescope and the plausibility of the discoveries, which

---

[70] Randall, *School of Padua*, 72–73.

he later confirmed when he put his hands on a telescope. But it was particularly to the support given by the Jesuit scientists at the Roman College that Galileo owed his acceptance. They had enormous prestige in the Catholic world, and by the end of October 1610 could visibly confirm Galileo's discoveries. If Galileo was to advance himself and his cause, the best way to do that was through gaining the support of Rome and the Roman intellectual world. A visit to the Eternal City was essential.

> The city and court of Rome are presently at the height of their greatness and prosperity, as anyone living there may see through clamorous examples of such pomp and splendour; and since, in this age, every province and court, even in the most far-flung and barbarous nations, is steeped in luxury and pleasure; so, in the city, and court of Rome, such manifestations are especially prevalent, and proliferate, since here, among the greater part of the most distinguished persons, the desire to live in great magnificence, with every convenience, is conjoined to the excessive wealth required to do so. Whence there is now the most excessive spending: and that lavish living which in other ages was the prerogative of a few of the foremost cardinals, now leaves its mark upon the magnificence of the buildings and the rich and noble decoration of the Palazzi, with truly regal pomp. Indeed, in these latter years as many public and private buildings, churches and palaces, streets and fountains have been made, that they alone would suffice to adorn a noble city.[71]

So the Venetian ambassador Paolo Paruta described the city in 1595. And when Galileo entered the city on March 29 1611, in a horse-drawn litter supplied by Cosimo, it was even truer. The Rome of 1611—the fifth largest city in Italy, after Naples, Milan, Venice, and Palermo—was very different from that of his last visit. While it continued to have a significant underclass, lacking as it did any real industry but the Church, with vagrants and beggars everywhere, a significant problem to public order, it had also become a much more magnificent and a much more clearly Christian city, endowed with new buildings and restored old ones, with painters and artisans flocking to the city as well as tourists and pilgrims in ever larger numbers. When Montaigne visited the city in 1580 it seemed more of a ruin,

[71] Helen Langdon, *Caravaggio: A Life* (New York: Farrar, Straus and Giroux, 1998), 43–44.

with two-thirds of it empty and the remnants of its classical past broken, scattered, or still buried—more like a medieval rabbit warren of a city than the city we know today. Sixtus V (1585–1590) had begun to transform it, laying out a new street plan and remodeling the city extensively, building an aqueduct (the Acqua Felice) to provide new water supplies, placing ancient obelisks in key places, building a new wing to the Vatican Library, rebuilding the Lateran palace and finishing the Dome of St. Peter's. Under Paul V, St. Peter's nave and façade would be finished (with Paul V's name prominently on it) and the nave of Old St. Peter's would be demolished.

The Borghese papacy also ushered in a new artistic style away from the somber simplicity and austerity of earlier Counter-Reformation popes. Under Pope Clement VIII naked saints were removed from Roman churches, and nude statues covered up, but under the Borghese a more exuberant and sensual style came to the fore, appealing directly to the senses, with greater color and emotion, with greater movement and intensity. It was the beginning of the Baroque, whose Roman apotheosis would be under Gian Lorenzo Bernini (1598–1680), whose family moved to Rome just five years before Galileo's triumphant visit, his Tuscan sculptor father Pietro having been commissioned to create a large relief of the Assumption of Mary for the façade of the Borghese Chapel in Santa Maria Maggiore. The story is told that shortly after the Bernini family arrived in Rome, Paul V himself predicted Gian Lorenzo's glorious future after asking him to do a drawing for him: "The young Bernini finished the drawing with such mastery that the pope stood there in admiration and was moved to simply exclaim to several cardinals who happened to be present on this occasion, 'This child will be the Michaelangelo of his age.' "[72] While in his childhood, Bernini trained under his father and Galileo's good friend the painter Cigoli, who was also working in the Borghese Chapel, and by his early twenties he was the dominant sculptor in Rome. He would soon become the great impresario of papal Rome.

The dominant figure at the court of Paul V, though clearly subordinate to his taciturn and imperious uncle, was his sister's son, whom we know as Scipione Cardinal Borghese (1577–1633). Ludwig von

---

[72] Franco Mormando, *Bernini: His Life and His Rome* (Chicago: University of Chicago Press, 2011), 13.

Pastor (1854–1928), the very proper German Catholic historian of the popes, describes him as "a stately man", genial, obliging, and cheerful, who "possessed an attractive manner, uncommon versatility and ability and a vivacious if not profound mentality". But even he admitted him to be "clever and worldly-wise", living on a "magnificent scale", more like a person of the world than a Prince of the Church, and whose "extraordinarily sumptuous" banquets were the talk of the town.[73] Made a cardinal at twenty-seven and showered with ecclesiastical offices and revenues, this ungainly and obese man had a great passion for art and an immense income with which to pursue it. While individuals could no longer accumulate bishoprics and other wealthy benefices since the Council of Trent, they could still receive multiple pensions from these (the costs coming from the benefices themselves), and wealthy monasteries could still be given *in commendam* to clerics who were not even monks there. Scipione possessed dozens of such monasteries and received a vast number of such pensions, besides receiving various other lucrative privileges and outright cash gifts (some 113,345 scudi) from the papal treasury.[74] And this was just what he received from the papal largesse, and did not include the pensions and gifts from other interested parties. Scipione grew obscenely rich: his average annual income from 1605, with the election of his uncle, until 1633, the year he died, was over 200,000 scudi, and in 1613 and 1614 it reached, respectively, 666,534 and 755,560 scudi. At a time when a family of five in Rome could live modestly on 90 scudi a year, these were extraordinary sums, especially in a Rome of such endemic poverty.[75]

He served as his uncle's "cardinal-nephew", a position that made him the pope's chief lieutenant, confidant, most intimate advisor, and *alter ego*. He presided over the Roman Curia, coordinated its activities, read the official correspondence of the papal nuncios and provincial and local governors of the Papal States, and interpreted

[73] Ludwig Pastor, *The History of the Popes from the Middle Ages*, trans. Ernst Graf, vol. 25 (St. Louis: B. Herder, 1937), 56, 63–64.

[74] Wolfgang Reinhard, "Papal Power and Family Strategy in the Sixteenth and Seventeenth Centuries", in *Princes, Patronage and the Nobility: The Court at the Beginning of the Modern Age, c. 1450–1650*, ed. Ronald G. Asch and Adolf M. Birke (Oxford: Oxford University Press, 1991), 336–37.

[75] Mormando, *Bernini*, 47 and xvii.

the pope's will to the outside world. He also served as the leader of the pope's faction, that network of clients and friends who owed their allegiance to the pope's family and upon which the family's power and success depended. At a time when power in all European courts was manifested in factions, power in the papal court was also manifested so.[76] There were factions belonging to the great papal families and other powerful Italian families, and factions connected to the various political powers such as Spain (the most influential of the powers) or France—and overlap was quite common. An enormous amount of attention had to be devoted to the building and maintaining of one's faction, with the dispensing of favors, gifts, titles, and offices, and with the promotion and advancement of friends and followers. For the papal faction, the selection of cardinals was particularly important since not only were they the greatest of all the prizes possessed by the papal family, but the long-term success of the family depended upon surviving the death of the papal uncle and electing a friendly successor, something for which loyal cardinals were essential.

Besides Cardinal Scipione many other family members benefited from the family's rise. Francesco, the pope's childless brother, was made general of the Church and admiral of the papal galleys; Giovanni Battista, another brother, was made commander of the papal guards, castellan of Sant' Angelo and the fortress of Ancona, and governor of the Borgo, the quarter between the Tiber and St. Peter's not belonging to the Roman commune. All these positions his son, Marcantonio, the family's sole heir, was to inherit. And to these offices, which had their own monetary perks and privileges, were added various other gifts, annuities, and outright cash payments (excluding the cardinal-nephew, the pope's family received cash payments of 645,950 scudi in his reign) with the ultimate the goal of transferring as much of this wealth as possible to the family permanently, especially by buying land and acquiring titles—and all the better if the land and titles were outside the Papal States and the vagaries of a new pope. The final step in this upward progress was to marry into the old Roman aristocracy, or better; and in 1619

[76] Maria Antonietta Visceglia, "Factions in the Sacred College in the Sixteenth and Seventeenth Centuries", in *Court and Politics in Papal Rome 1492–1700*, ed. Gianvittorio Signoretto and Maria Antonietta Visceglia (Cambridge: Cambridge University Press, 2002), 99–131.

Marcantonio married Camilla Orsini, the daughter of the Duke of Bracciano and member of the ancient Orsini clan of the Roman nobility. In 1616 Marcantonio had been made prince of Sulmona in the Spanish-controlled Kingdom of Naples and in 1620 (for the price of appointing the ten-year-old prince Fernando, son of King Phillip III, a cardinal and administrator of the archbishopric of Toledo, the wealthiest benefice in Christendom) Marcantonio received the rank of *grande de España* (grandee of Spain), the highest rank of the Spanish nobility, something rare among the Roman nobility.[77]

Among the decrees of the last session of the Council of Trent in 1563 was one enjoining on the bishops and cardinals a modest and frugal life and forbidding them to enrich their relatives or familiars with the revenues of the Church. Despite this, such nepotism and patronage continued among the bishops and cardinals. It also continued among the popes. It was expected that one's family and familiars would play a part and share in the benefits of any pontificate. This was not just because they would be the most loyal and dependable, and would strengthen the pope's position, but also because it would be wrong not to do so. It would be a lack of *pietas*, of duty and devotion to one's friends and family to whom one owes so much. The great Carlo Borromeo, the model Catholic bishop, had risen through nepotism, being the nephew to Pope Pius IV. While nepotism was expected, it also had limits. Under Clement VIII Aldobrandini (r. 1592–1605), it started to get out of control again, with Clement making cardinals of two of his nephews, Pietro and Cinzio Aldobrandini, and later of his sixteen-year-old grandnephew Silvestro. He also endowed all of them, and other family members, with every sort of office, title, fief, and income possible, making the Aldobrandini a truly princely house.

The Borghese, who became the great rivals of the Aldobrandini, were to outdo them. If Pietro Cardinal Aldobrandini could build his spectacular Villa Aldobrandini in Frascati twelve miles outside of Rome, with its splendid art collection, its magnificent park, and water theater, then Scipione Cardinal Borghese would do as much and more with his Villa Borghese just outside Rome. Cardinal Borghese would also be a great patron of Bernini, who produced his

[77] Reinhard, "Papal Power and Family Strategy", 339.

statues *Aeneas and Anchises*, *Persephone*, *Apollo and Daphne*, and *David* for him, now to be found in the Villa Borghese, which he built to showcase his art collection. Among those working for the Borghese was Galileo's friend Lodovico Cardi (1559–1613), known as Il Cigoli, who, at the very time of his visit, was painting a fresco in the cupola of the Pauline (Borghese) Chapel of Santa Maria Maggiore, *The Immacolata*, where the Virgin is standing atop a pockmarked moon, much like Galileo's moon, instead of a perfectly smooth, spherical orb as had been the norm.

Scipione's passion for art was such that nothing could stand in his way, and as cardinal-nephew there was little that could stop him. He stole Raphael's *Deposition* from the Baglioni family chapel in Perugia. He arrested Domenico Zampieri (also called Domenichino, 1581–1641) to make him sell his *The Hunt of Diana* to him. He imprisoned the very prominent and respected Roman painter Cavalier d'Arpino (Giuseppe Cesari, 1568–1640), the favored painter of Clement VIII, on the charge of possessing illegal weapons, so that he could confiscate more than a hundred paintings from his collection. He bought seventy-one paintings from Paolo Cardinal Sfrondati (1560–1618), who was the cardinal-nephew of Gregory XIV, but never paid him. When the brilliant but turbulent painter Caravaggio (1571–1610), having killed a man in a swordfight and fled Rome, wished to be pardoned and allowed to return, Scipione agreed but only if he received all of his unsold canvases.[78]

While Paul V's personal conduct had always been exemplary and above suspicion, and he was even reputed a virgin, his nephew's personal life was unsavory. His means of acquiring real estate (and the Borghese acquired a lot of it in the Papal States) was to sign a legal purchase contract but never to pay the money owed despite reminders. His relationship with his majordomo Stefano Pignatelli became such a source of scandal that Paul V was forced to banish Pignatelli from Rome to protect his nephew's reputation. Scipione fell into such a depression that the pope was forced to call him back. Later when Pignatelli was made a cardinal a pasquinade (after the ancient statue, Pasquino, where Romans attached their anonymous

[78] Andrew Graham-Dixon, *Caravaggio: A Life Sacred and Profane* (New York: W. Norton & Company, 2010), 426–37.

satirical thoughts or criticisms) was posted that read: "Why is every-one so surprised? Spain campaigns for her candidates, France for hers; everyone wants his own man to be made cardinal. So why shouldn't Cardinal Scipione's penis get what it wants too, its own man in the College of Cardinals?"[79]

> Thus I came into Bellermines chamber that I might see this man so famous for his learning and so great a Champion of the Popes: who seemed to me not above forty yeeres old, being leane of body, and something low of stature, with a long visage and a little sharpe beard uppon the chin, of a browne colour, and a countenance not very grave, and for his middle age, wanting the authority of grey heires. Being come into his chamber and having made profession of my great respect to him, I told him that I was a Frenchman, and came to Rome for performance of some religious vowes, and to see the monuments, especially those which were living, and among them himselfe most es-pecially, earnestly intreating, to the end I might from his side returne better instructed into my Countrey, that he would admit me at vacant houres to enjoy his grave conversation. He gently answering, and with gravity not so much swallowing the praises I gave him, as shew-ing that my company should be most pleasing to him, commanded his Novice, that he should presently bring me in, when I should come to visit him, and so after some speeches of curtesie, he dismissed me, who meant nothing lesse then to come againe to him.[80]

Already by 1594, the time of the visit of Fynes Moryson described above, Bellarmine had already become a famous man, someone whom visitors to the Eternal City had to meet—even for a strong Protestant like Moryson though Bellarmine was not aware of that. He was also someone whose support was extremely important to have. This was even truer later on after he became a leading cardinal in the Roman Curia, a cardinal inquisitor, and someone whom many considered a living saint. Galileo would not only visit him on his trip to Rome; he would even show him the heavens through his telescope.

Galileo would remain in the Eternal City for more than two months as an official guest of the Grand Duke at the Palazzo Firenze near the

[79] Mormando, *Bernini*, 46–50.
[80] Moryson, *Itinerary*, 1:304.

Pantheon, the Tuscan embassy in Rome; and the next day after his arrival he called on Cardinal del Monte, the Medici agent at the papal court, and went to the Roman College to meet Christopher Clavius, and the other Jesuit scientists Christopher Grienberger (1561–1636), Odo van Maelcote (1572–1615), and Giovanni Paolo Lembo (1570–1618). Over the next two months he had an audience with Pope Paul V, met or made visits to the leading prelates and princes in Rome, including Cardinal Bellarmine, and displayed his telescope at a number of high society soirees. But the highlight of his visit to Rome was the celebration held for him at the Roman College. Besides its faculty and students, and the scholarly community of Rome, there were any number of prelates and noblemen, at least three cardinals, with the highly laudatory speech by Maelcote.

Shortly before Galileo was fêted at the Roman College, Bellarmine (April 18, 1611) sent a memorandum to Clavius and the Jesuit scientists about Galileo's findings: asking whether they agreed with him concerning the multitude of fixed stars in the heavens, the number of the stars, the satellites of Jupiter, the phases of Venus, the shape of Saturn, and the surface of the moon. Maelcote replied for all within five days saying that they agreed with Galileo on the first four but concerning the nature of the surface of moon there was no consensus. Clavius rejected it as rough and covered with craters, claiming that it only appeared as such because the moon had rarer and denser parts and was covered by a transparent mantle like a smooth glass envelope.

While Galileo was in Rome, he was also inducted into one of the first scientific academies in Europe, the *Accademia dei Lincei* (Academy of Lynxes), founded by Prince Federico Cesi (1585–1630) in 1603 with three other friends. The Lynceans, however, were not the first, for there had been an earlier scientific academy in Naples, founded in 1560 by the brilliant polymath Giovanni Battista della Porta (1535–1615): the *Accademia dei Segreti* (*Academia Secretorum Naturae*). This Academy of the Secrets of Nature, which met at della Porta's home, had as a condition of membership that candidates should each present a new fact in natural science. His many interests included magic (natural magic) and other occult sciences, and the academy disbanded in 1580 after he had been brought before the Inquisition. Della Porta met Cesi in 1603, and when he joined the Lynceans in July 1610, his European-wide fame added great prestige and authority to a rather unknown group. The sharp-eyed Lynx, the symbol of the Lynceans, was itself taken from the title page of della Porta's *Natural Magic* (final edition in 1589). Cesi's goal was to organize, collaboratively, scientific research of all the natural sciences; and while in its early years its membership was only a few dozen (and its goal was to create a network of scholars all across Europe and beyond), it published important works in botany (Cesi's real interest), zoology, geology, and astronomy. By his wealth, diplomatic skill, and good connections, Cesi was able to bring together an organization not only free of any Aristotelian presuppositions but one that could challenge the Jesuit hegemony on science. Galileo, the sixth person to join, was always proud of his membership in the Lynceans, and he displayed this title on his books; and they would publish his *Letters on Sunspots* (1613) and his *Assayer* (1623).

While Galileo's visit to the Eternal City was for the most part a triumph, there were already hints of a darker future. Not all were won over, and at a meeting of the Inquisition on May 17, with Bellarmine attending, it was asked if Galileo was mentioned in any of the investigations of Cremonini. It is very plausible that Bellarmine was the origin of the query.[81] And in a much later official letter of the Tuscan ambassador Piero Guicciardini, who took over as ambassador a few weeks into Galileo's 1611 visit to Rome, he wrote (with the most sensitive parts in cipher):

> His teaching, and something else, was not to the taste of the Advisors and the Cardinals of the Holy Office. Among others, Bellarmine told me that, however great their respect for the Grand Duke, if Galileo had stayed here too long, they could not have avoided looking into the matter. I gave Galileo a hint or a warning since he was staying here, but I fear that it did not give him great pleasure.[82]

---

[81] Richard S. Westfall, "Science and Patronage: Galileo and the Telescope", *Isis* 76 (1985): 11.

[82] William Shea and Mariano Artigas, *Galileo in Rome: The Rise and Fall of a Troublesome Genius* (Oxford: Oxford University Press, 2003), 47.

Chapter Three

Urania's Children: Science, Astronomy, and Astrology from Antiquity to the Renaissance

In the Middle Ages and Renaissance, science (what they would have called natural philosophy) was profoundly Aristotelian. For Dante, Aristotle (384–322 B.C.) was "the Master of those who know", and Saint Thomas Aquinas, considered the greatest Doctor of the Church and the standard by which Catholic theology is measured, called him "the Philosopher". He wrote so extensively and so systematically that there was little he did not write on. Father of biology and inventor of formal, syllogistic logic; systematizer of poetics and political science; his "natural books" (*Physics*, *On the Heavens*, *On the Soul*, *On Generation and Corruption*, *Meterology*, and the *Parva naturalia* or *Short Physical Treatises*) were the basis of natural philosophy in the medieval and Renaissance universities and would set the stage for the West's understanding of the physical universe up into the seventeenth century. His cosmos (and he believed that the existence of any other worlds was impossible) was a great sphere that had no beginning and would have no end, with no empty spaces, in a single system of concentric, invisible, transparent spheres which held within the shell of their inner and outer walls the individual planets. In this universe lay two major divisions of terrestrial and celestial regions, the dividing line being the curved inward surface of the sphere of the moon. Beneath the moon lay the terrestrial region, the world of the four elements of earth, water, air, and fire, or some combination thereof, a world of constant change, of bodies coming to be and passing away. The celestial region, the realm of the other planets and fixed stars, was unchanging, the world of the fifth element of ether (or quintessence), an incorruptible and eternal substance from which all the planets and stars were composed. This realm experienced no change except of place, making it nobler than the terrestrial region, as its planets and stars moved in an eternally uniform circular motion. Below the moon, at the center of this cosmos, was the unmoving sphere called the earth, the heaviest and lowest part of the universe (the gutter of the universe) and for that reason greatly influenced by the far superior celestial realm.

While Aristotle posited a God—this Unmoved Mover beyond the sphere of the fixed stars at the outermost circumference of the cosmos, who was eternal and incorruptible, unchanging and immobile and perfect—it was quite unlike the God of the Christians. This God did not create the world and had no interest in it. He did not know of the existence of individual human beings, nor intervene on their behalf. He neither offered paradise nor threatened with hell, absorbed in his own thoughts since thought was the highest of all activities and his thought was the highest of all kinds of thought. Nor was he the only God, but the first among equals since each sphere, and there were fifty-five of them in Aristotle, had its own immaterial unmoved mover (or intelligence).

The universe that Christendom inherited, however, was really Aristotelian and Ptolemaic, a combination of Aristotle and Ptolemy, though Aristotle was clearly dominant. Claudius Ptolemy (c. A.D. 90–c. 168) was a Greek-speaking Egyptian who lived and died in Alexandria during the Roman Empire, at that very time which Gibbon considered that the condition of the human race to be at its most happy and prosperous. A mathematician, astronomer, astrologer, and geographer, affiliated with the great Museum (Musaeum, "Institution of the Muses", the great research center founded by the Ptolemies in Alexandria) and its library, he was the author of the most comprehensive treatise on astronomy from antiquity, the *Mathematical Syntaxis*, known to us as the *Almagest* (from the Arabic for "the greatest") and a companion volume on astrology most commonly called the *Tetrabiblos* (Four books). While there were similarities between the cosmologies of Aristotle and Ptolemy (for example, the earth is at the center in each), they were fundamentally incompatible. While Aristotle may have been "the Master of those who know" and he discussed many things, he did not discuss mathematics and astronomy (which have their own traditions separate from Aristotle), and his description of how the stars should work did not match how they did work. In Aristotle the motion of the celestial bodies was circular and regular around a fixed point of the earth, but since his concentric spheres could not account for observed variations in the distances of the planets from the earth (as seen in their variations of brightness) and in their retrograde motions (their appearing to stop and even go backward), the Greek mathematical astronomers culminating with Ptolemy created epicycles, eccentrics, equants, and deferents, a complex series of mathematical mechanisms to keep the observational data and Aristotelian worldview in sync, "to save the phenomena", as the phrase goes. A compromise was set up whereby Aristotle's natural philosophy described the world and supplied the reasons for celestial motion while astronomy and mathematics (which did not belong to natural philosophy but to a lesser science) calculated the paths of the heavenly bodies. It worked only because scholars saw the mathematical techniques as fictions providing useful information but which did not truly describe reality.

Such then was the Aristotelian-Ptolemaic world: an earth-centered world encircled, in order, by the moon, Mercury, Venus, the sun, Mars, Jupiter, Saturn, the fixed stars, and the Sphere of Prime Mover. Aristarchus of Samos (310 B.C.–c. 230 B.C.) proposed a model of the universe where the sun was at the center with the earth revolving around it (a heliocentric theory as we see in Copernicus), and Seleucus of Seleucia (c. 190 B.C.–fl. 150s B.C.) later championed his view, but this view was rejected by Ptolemy and all other astronomers. The Pythagoreans had a slightly different view where all the planets, including the sun, revolved around a central fire. Thus, when Galileo described the heliocentric universe in the seventeenth century, as in his *Two Chief World Systems*, he called it, wrongly, Pythagorean.

Greek science flourished in the Roman Empire (Ptolemy himself bore a Roman name and was a Roman citizen), even if its study attracted few people and had little influence. Nor was a purely rationalist or scientific mind-set the cultural norm. Appeals to religion, traditional as well as to newer movements, appeals to superstition or the stars, and appeals even to magic and the occult were far more common and significant. Even philosophy, except for the Epicureans, as we mentioned, generally had a religious element which became even more intensely and explicitly religious in late antiquity. It even included a strange element of theurgy (ritual magic and divination) and bloody animal sacrifice among some later Neoplatonists. Gilbert Murray (1866–1957), the eminent Australian-born British classical scholar, looking at the Greek mind after the death of Demosthenes (322 B.C.), in the period we call Hellenistic, saw there a profound decline, a rise of asceticism, mysticism, and pessimism, "a loss of self-confidence, of hope in this life and of faith in normal human effort, of a despair of patient inquiry and a cry for infallible revelation", which he blamed on "a failure of nerve". But in this he was, to a great extent, reflecting his own idealized view of the Greek way and his own distaste for all religion, and for Christianity in particular. For him, the classical world "mistrusting Reason, wary of arguments and wonder, flung itself passionately under the spell of a system of authoritative Revelation, which acknowledged no truth outside itself and stamped free inquiry as sin."[1]

The Greek mind was never as secular and rationalistic as he imagined, nor, as we will see, was the Christian mind so opposed to reason and argument. That the intellectual vigor and confidence of the Greeks and Romans waxed and waned there can be no doubt, as it waxes and wanes in our own civilization depending upon the successes or failures that it experiences; that among the elites (and our evidence fragmentary as it is comes primarily from here) there were periods of greater religious belief and otherworldliness is certainly true, as it was true of late antiquity. But in every generation man has been a plaything of fickle chance (*tychē*) or inexorable fate (*heimarmenē*). It was not a decadent Hellenistic poet of some forlorn Greek colonial outpost who wrote, "I am a child of Chance," but Sophocles in a play (*Oedipus Rex*) presented at the very apex of Athenian greatness, and in a play that is a reassertion of the religious view of a divinely ordered universe.[2]

From the rise of science with the pre-Socratics until the beginning of the Christian era, there have been only two known instances of a clash between science and religion.[3] The teacher and friend of Pericles, the philosopher Anaxagoras of Clazomenae (c. 500–428 B.C.), was charged with impiety (atheism) and put on trial because he believed the sun to be just a fiery hot mass of metal—though his

---

[1] Gilbert Murray, *Five Stages of Greek Religion* (Boston: Beacon Press, 1951), 123; and Gilbert Murray, *A History of Ancient Greek Literature* (New York: D. Appleton, 1916), 404.

[2] Bernard Knox, *Oedipus at Thebes* (New Haven, Conn.: Yale University Press, 1957), 47.

[3] Edward Grant, *Science and Religion, 400 B.C. to A.D.: From Aristotle to Copernicus* (Westport, Conn.: Greenwood Press, 2004), 15–16.

friendship with Pericles was very likely the real cause of the charge. He was banished, or forced to flee, or released and allowed to leave (the sources say different things), and perhaps even committed suicide because of the humiliation. The other case was that of Cleanthes (c. 330–230 B.C.), the second head of the Stoic school in Athens, who charged Aristarchus with impiety because he moved the earth from the center of the universe and set it in motion. However, we know of no formal charge being brought and no consequences that followed. While religion in the ancient world was neither very dogmatic nor very organized, a charge of atheism could still become, as Socrates and the early Christians knew, quite deadly.

Christianity could only have been seen as a bizarre religion to a great many of the educated pagans of the Roman Empire, particularly to the philosophically minded. It was not really its monotheism that was the difficulty for many of them, for, despite their public adhesion to the anthropomorphic gods of the Greco-Roman world and involvement in the gods' cultic manifestations which was part and parcel of public life, there was also a trend toward a kind of monotheism at this time, of a single ultimate controller and source of the world, whether it be the Unmoved Mover of Aristotle, the Universal Reason or *Logos* of the Stoics, or the One of the later Platonists. However, none of these really describes the Christian God since they are not personal or at least personally involved. Rather, they are something universal, immutable, self-contained, without desire or interest in human affairs, and as Funkenstein put it so well, the idea that God should choose some people over others (e.g., the Jews or the Christians) or be active in history, "an all-powerful busybody", "insulted the Greek sense of harmony".[4] The Christian God not only took on a physical human nature, "the *Logos* became flesh", but he was also crucified, died, rose again, and even now dwells in a transfigured but very real body. And to top it off, somehow this single God is also triune, a Trinity of Father, Son, and Holy Spirit. The contrast between the God of the philosophers and the God of the Christians (or as the great scientist Blaise Pascal would put it, the God of Abraham, Isaac, and Jacob) could not be more different, and the tension between the two was inevitable.

---

[4] Amos Funkenstein, *Theology and the Scientific Imagination: From the Middle Ages to the Seventeenth Century* (Princeton, N.J.: Princeton University Press, 1986), 125.

The Christian response to pagan learning was complex and varied. The same Saint Paul who wrote to the Corinthians that the wisdom of this world is foolishness to God (1 Cor 3:19) and had warned the Colossians against being deceived by any empty, seductive philosophy that follows mere human traditions and encouraged them to live by faith (Col 2:7–8) also disputed with Epicurean and Stoic philosophers in Athens, the intellectual and cultural center of the pagan Greek world, and preached on its venerable Areopagus in a rather philosophical way (Acts 17:16–33), using largely Stoic arguments, even quoting a verse of the Greek pagan poet Aratus.[5] While he was skilled in the rabbinic learning of the Scriptures, he was also a second-generation Roman citizen (a significant privilege particularly for a provincial Jew) who had received a good education of a classical Greco-Roman sort and could utilize its rhetorical and philosophical styles.

The Roman theologian Hippolytus (A.D. 170–235) saw pagan philosophy as the origin of all heresies, and there was some justified fear of the Christian message being absorbed by more popular alien systems, with Gnosticism being a particular danger. Gnosticism, from the Greek *gnosis* for "knowledge", was a collection of religious movements which fused elements of Greek philosophy, Judaism, Christianity, magic, and much else into a mélange of interpretive myths offering a secret knowledge that would free the individual from the evil prison of this material world. It was opposed not only by many Christian authors but also by such important pagan philosophers as the Neoplatonist Plotinus (c. A.D. 204–270). In his tractate "Against the Gnostics" from his great work the *Enneads*, he describes them as corruptors of the Platonic tradition, as immoral, dangerous, and irrational, and as blasphemers against the Creator and his cosmos for affirming them as evil.

Most famous of those Christian thinkers who opposed all pagan knowledge was the North African Latin theologian Tertullian (c. A.D. 160–c. 225), who famously wrote: "What indeed has Athens to do with Jerusalem? What concord is there between the Academy and the Church?... Away with all attempts to produce a mottled Christianity

---

[5] Werner Jaeger, *Early Christianity and Greek Paideia* (Cambridge, Mass.: Belknap Press of Harvard University Press, 1961), 11.

of Stoic, Platonic and dialectic composition!"[6] However, this same thinker who railed against Plato, Aristotle, and philosophy in general, and who saw pagan philosophy as demonic and a source of heresy, himself used philosophical terminology to help illuminate the Trinity and other Christian doctrines. He also believed that the soul was by nature Christian (*anima naturaliter Christiana*) and that each had the capacity and need to pursue God by reason as well as by faith when he wrote, in a passage which will be quoted by Galileo in his important discussion of Scripture interpretation, the *Letter to the Grand Duchess Christina*, "We postulate that God ought first to be known by nature, and afterward further known by doctrine—by nature through his works, by doctrine through official teaching."[7]

More common was a positive view of pagan philosophy, of men such as Justin Martyr (A.D. 100–165), Clement of Alexandria (c. A.D. 150–c. 219), both converts, and Origen (c. A.D. 184–c. 254), of devout Christian parents and whose father died a martyr.

Justin Martyr, born of Greek parents near Samaria in Palestine, moved through various teachers of philosophy, first a Stoic, then an Aristotelian, then a Pythagorean, and finally a Platonist with whom he seemed to have found true wisdom. One day, however, he met an old man on the seashore who told him that the Jewish prophets and their writings were more reliable than the reasoning of the philosophers. These miracle-working prophets were inspired by God and spoke the truth about him and his son, Jesus Christ. Justin was moved by his words, and soon Christianity became his "true philosophy", leading him even to open a school in Rome, where he taught it freely and where he was martyred by beheading during the reign of the philosopher-emperor Marcus Aurelius. Justin had a great though not uncritical appreciation for the achievements of Greek philosophy. He considered Socrates a Christian *avant la lettre*, believing that rational creatures, insofar as they are rational, share in the *Logos*, the creative reason that ordered the cosmos and became incarnate in Jesus Christ.

For Clement of Alexandria both Greek philosophy and the Old Testament were tutors which led people to Christ, the Christ who

[6] Tertullian, *On the Prescription of Heretics* 7, trans. T. Herbert Bindley (London: SPCK, 1914).
[7] Tertullian, *Against Marcion* 1.18.

sums up all wisdom and who is the full truth only partially seen in the different philosophical schools. He saw the usefulness of Greek philosophy both as a preparatory study and as an aid, and he brought into Christian discourse (from the Jewish Philo of Alexandria) the maxim that "philosophy was the handmaiden of theology."[8]

Origen, to whom so much of Christian theology is in debt and of whom the great Renaissance humanist Erasmus once wrote that he learned more of Christian philosophy from one page of him than from ten pages of Saint Augustine, was the onetime head of the catechetical school at Alexandria. He was a Scripture scholar par excellence and extremely well-read in Greek philosophy; and while he could be sharply critical of aspects of it, he nonetheless taught his students the full range of pagan knowledge ("the spoils of the Egyptians") as is recounted in this discourse of his onetime student and future bishop Gregory the Wonderworker (A.D. c. 213–c. 270), who was won from the study of law by Origen's irresistible charm and persuasiveness.

> He required us to study philosophy by reading all the existing writings of the ancients, both philosophers and religious poets, taking every care not to put aside or reject any ..., apart from the writings of atheists.... He selected everything that was useful and true in each philosopher and set it before us, but condemned what was false.... For us there was nothing forbidden, nothing hidden, nothing inaccessible. We were allowed to learn every doctrine, non-Greek and Greek, both spiritual and secular, both divine and human; with the utmost freedom we went into everything and examined it thoroughly, taking our fill of and enjoying the pleasures of the soul.[9]

In the Latin West the authority of Saint Augustine of Hippo (A.D. 354–430) was of particular significance, and all Western theology (Catholic and Protestant) is profoundly influenced by him. Son of a mixed marriage (his father was pagan and his mother Christian),

---

[8] Henry Chadwick, *Early Christian Thought and the Classical Tradition: Studies in Justin, Clement and Origen* (Oxford: Clarendon Press, 1966), 9–22, 31–65.

[9] David C. Lindberg, "Science and the Early Church", in *God & Nature: Historical Essays on the Encounter between Christianity and Science* (Berkeley: University of California Press, 1986), p. 24. See Chadwick, *Early Christian Thought*, 66–123.

he was a member of the curial class and therefore an inheritor of the traditional upper-class rhetorical culture of the Greco-Roman world. As anyone can know from reading his classic, the *Confessions* (A.D. 398), after reading Cicero's *Hortensius*, his whole life became a pursuit of true wisdom. He rose to become the professor of rhetoric at Milan, then the residence of the emperor in the West, and was baptized in 387 in Milan by Saint Ambrose (A.D. 337–397), after having gone through Manichaeanism (a variety of Gnosticism), the skepticism of the New Academy, and then the Neoplatonism of men like Plotinus and Marius Victorinus, an eminent but somewhat older contemporary Roman rhetorician and philosopher whose conversion Augustine also recounts in the *Confessions*. He was very conversant with the science of his day, and one element in his disillusionment with Manichaeanism was its ignorance of science and the ridiculousness of its cosmological descriptions. In his *De Genesi ad Litteram* (*Literal Commentary on Genesis*) he discussed how to interpret Genesis, and Scripture in general, in reference to the science of the day, and passages such as these were used very effectively by Galileo in his *Letter to the Grand Duchess Christina*.

Usually, even a non-Christian knows something about the earth, the heavens, and the other elements of this world, about the motion and orbit of the stars and even their sizes and relative positions, about the predictable eclipses of the sun and moon, the cycles of the years and the seasons, about the kinds of animals, shrubs, stones, and so forth, and this knowledge he holds to as being certain from reason and experience. Now it is a disgraceful and dangerous thing for an infidel to hear a Christian, presumably giving the meaning of Holy Scripture, talking nonsense on these topics, and we should take all means to prevent such an embarrassing situation, in which people show up vast ignorance in a Christian and laugh it to scorn. The shame is not so much that an ignorant individual is derided, but that people outside the household of the faith think our sacred writers held such opinions, and, to the great loss of those for whose salvation we toil, the writers of our Scripture are criticized and rejected as unlearned men. If they find a Christian mistaken in a field which they themselves know well and hear him maintaining foolish opinions about our books, how are they going to believe those books in matters concerning the resurrection of the dead, the hope of eternal life, and the kingdom of heaven, when they think their pages are full of falsehoods on facts which they themselves have learnt from experience and the light of reason?[10]

When asked about what is the belief of Scripture in reference to the form and shape of the heavens, Augustine replied that the sacred writers did not concern themselves with such topics which have no bearing on salvation, which, in any case, are a waste of time since they do not pertain to eternal happiness and would take up precious time from what was spiritually beneficial. "But someone may ask: Is not Scripture opposed to those who hold that heaven is spherical, when it says, *who stretches out heaven like a skin?*" His reply was that if they could establish their doctrine with undeniable proofs, then one would have to show how Scripture was not opposed to the truth of their conclusions but reconcilable.[11] In fact, despite what Scripture described, Augustine, and other Christian commentators, generally accepted the view of the cosmos that was part of the intellectual framework of their time and reflected the science of the day. At the end of the second book of his *Literal Commentary on Genesis* he left some sage advice.

> Meanwhile we should always observe that restraint that is proper to a devout and serious person and on an obscure question entertain no rash belief. Otherwise, if evidence later reveals the explanation we are likely to despise it because of our attachment to our error, even though this explanation may not be in any way opposed to the sacred writings of the Old or New Testament.[12]

As we see, Augustine was well aware that not all believers used their minds well, and growing up among Christians in Roman North Africa, he had experienced an often backward and even fundamentalist Christianity. It was Bishop Ambrose of Milan who had shown him that Scripture could often be read in a figurative and not literal way, and it was this which opened his path to conversion. For Augustine,

---

[10] Saint Augustine, *The Literal Meaning of Genesis*, trans. John Hammond Taylor, 2 vols. Ancient Christian Writers (New York: Paulist Press, 1982), 1:19, 42–43.

[11] Ibid., 2:9, 59.

[12] Ibid., 2:18.

while faith and revelation were the best means to achieving truth, and Scripture would be given preference where there were no demonstrative arguments against it, reason was not to be ignored and philosophy was still the handmaid of theology, useful tools that helped us to our supernatural end.

Nor was Eastern Christianity despite its great intellectual figures altogether free of this fundamentalist temptation as we see in the Anthropomorphite Controversy. In A.D. 399 Bishop Theophilus of Alexandria issued his annual festal letter announcing the dates of Lent and Easter. But it also included an aggressive attack on those who believed that God had a bodily form, calling it a heresy. (At this time Theophilus was attempting to demolish Egyptian paganism and its idol worship, as we see in his destruction of the Sarapeum in Alexandria, and so was stressing the intrinsic incorporeality of God.) When the letter was read to the monks of the Egyptian desert, the new heroes of the Christian world, a vast majority opposed it violently as contrary to the plain teaching of Scripture. The bulk of the monks were simple peasants, and one old monk wept, "Woe is me! They have taken my God from me, and now I have no one to hold onto and I no longer know whom I should adore or address."[13] Many marched on Alexandria, rioting and threatening the bishop's life. Theophilus quickly backpedaled, affirmed God's corporeality, and instead condemned Origen (Origenism) as the source of this spiritualizing tendency, leading to the purging or flight of some three hundred monks, the intellectual elite, from the Egyptian monasteries and the more universal condemnation of this great Christian teacher.

While in the Latin-speaking provinces of the Roman Empire there was a network of schools of rhetoric which supplied higher education, there were no permanent schools of philosophy as we see in the Greek-speaking East, at Athens and Alexandria, and later at Constantinople, and perhaps at other large cities there. While most famous for his *Consolation of Philosophy*, a work he wrote awaiting his execution and a favorite right into the Renaissance, Boethius (c. A.D. 480–525/526), an aristocratic Christian Roman and philosopher who had served as the chief minister of the Ostrogothic ruler of Italy, was

---

[13] Paul A. Patterson, *Visions of Christ: The Anthropomorphite Controversy of 399 CE* (Tübingen: Mohr Siebeck, 2012), 4.

more important for his attempt to pass on the philosophic heritage of antiquity. He intended to translate all of Plato and Aristotle into Latin, but in the end he just translated five of Aristotle's logical works: the *Categories, On Interpretation, Sophistical Refutations, Prior Analytics,* and *Topics* (and probably the *Posterior Analytics*) and Porphyry's *Isagoge* (his introduction to Aristotle's logic), with commentaries on some of them, as well as writing his own treatises on logic, all of which formed with some other texts the "old logic" (*logica vetus*) which would play an important part in the revival of learning in the eleventh century. In his five theological tractates Boethius also applied this logical reasoning to theology, something which would be more aggressively done among the medieval scholastics, which is why the humanist Lorenzo Valla in the fifteenth century would call him the last of the Romans and the first of the scholastics.

In the Christian East, in what we now call the Byzantine Empire, classical learning and education continued even after the Roman collapse in the West in the fifth century. Emperor Justinian's closing in 529 of the Platonic Academy at Athens (Plato's Academy had closed much earlier, and this was a new creation from the turn of the fifth century) was not the death of classical philosophy as some would have it; and those seven pagan philosophers who left the empire for Persia in high dudgeon in 531 returned in 532 to live unmolested and productive lives.[14] The philosophical schools continued at Constantinople and at Alexandria, where the pagan Neoplatonic philosopher Olympiodorus continued to teach and write publicly into the 560s, with the school continuing until the Arab conquest of 641. The Alexandrian school under the pagan Ammonius (c. 440–c. 520) produced two of the most significant interpreters of Aristotle from late antiquity, both Neoplatonists, the pagan Simplicius (c. 490–c. 560) and the Christian John Philoponus (c. 490–c. 570). Philoponus wrote the most insightful and destructive criticism of Aristotle that the ancient world had ever seen, particularly opposing his views on the eternity of the world, that the heavens were made of an incorruptible separate

---

[14] Edward Watts, "Justinian, Malalas and the End of the Athenian Philosophical Teaching in A.D. 529", *Journal of Roman Studies* 94 (2004): 168–82; and Edward Watts, "Where to Live the Philosophical Life in the Sixth Century? Damascius, Simplicius and the Return from Persia", *Greek, Roman, and Byzantine Studies* 45 (2005): 285–315.

fifth element different from the four of the terrestrial world, and on motion. Most of his work would remain unknown in the West until the sixteenth century, though some of his ideas would come through William of Moerbeke's translation of Simplicius' commentary on Aristotle's *On the Heavens* (where he attacks Philoponus) and Moerbeke's partial translation of Philoponus' commentary on Aristotle's *On the Soul*. Galileo would mention him by name many times, and with respect, in his own early writings on motion.

> In a sermon on repentance preached at Santa Maria Novella in Florence on 23 February 1306, the Dominican Fra Giordano of Pisa, while providing our best evidence of the invention of eyeglasses in the 1280s, incidentally sang the praises of the recent invention of invention. "Not all the arts," he said, "have been found; we shall never see an end of finding them. Every day one could discover a new art ... indeed they are being found all the time. It is not twenty years since there was discovered the art of making spectacles which help you see so well, and which is one of the best and most necessary in the world. And that is such a short time ago that a new art, which had never before existed, was invented.... I myself saw the man who discovered and practiced it, and I talked with him."[15]

The American scholar Lynn White Jr. (1907–1987), to whom the above quotation belongs, was a major figure in the transformation of our understanding of the Middle Ages. In a paper he gave at the 1974 semicentennial of the History of Science Society, he remarked that the Middle Ages were not what they used to be. As an undergraduate at Stanford in the 1920s, he learned "two firm facts about medieval science: (1) there wasn't any, and (2) Roger Bacon was persecuted by the church for working at it."[16] While Catholics have generally had a profound appreciation of the Middle Ages (think of the popular and often-reprinted 1907 book, *The Thirteenth, The Greatest of Centuries*, by the American Catholic historian James J. Walsh), the rest of the

---

[15] Lynn White Jr., *Medieval Science and Technology: Collected Essays* (Los Angeles: University of California Press, 1978), 221.

[16] Lynn White Jr., "Introduction: The Study of Medieval Technology, 1924–1974: Personal Reflections", in *Medieval Science and Technology*, xi–xii.

world generally did not. It was due to individuals like White that we now have a far more accurate, and far more appreciative, view of that period. White was amazed by the amazing creativity and dynamism of the medieval West in the area of technology, even claiming that it had "invented invention"; that is to say, it had developed a mind-set and expectation of invention that was unique in human history. This same creativity and dynamism was true of intellectual life as well, and science would be one of it beneficiaries.

With the growth of stability in Europe after the crisis of the ninth and tenth centuries caused by the attacks of the Vikings, Magyars, and Arabs, education revived with major schools at certain monasteries and more importantly at cathedrals. It was from these cathedral and urban schools that we see the rise of the universities and scholasticism (the method of the schools) which used the dialectical and analytical tools of logic, reason, and disputation to investigate reality. This dialectic was combined with the use of authorities, the Bible, and the Fathers of the Church for theology and ancient pagan learning, particularly the works of Aristotle, for everything else. Aristotle had written a large number of dialogues and other popular treatises which were widely read and praised for their literary elegance, but none of these survived antiquity. Rather, what did survive were his lecture courses delivered in his school at Athens, the Lyceum, and kept in its library. These were finally edited and circulated in the first century B.C., but even then they were little utilized until the second century A.D. The Aristotelian Alexander of Aphrodisias commented on them, as did generations of Neoplatonists who saw Aristotle as a preparation for Plato and tended to harmonize him with Plato.

While some of Aristotle's works had been translated into Latin by Boethius, most of the West's knowledge of Greek philosophy and science was only known through the Church Fathers or through Latin handbooks and encyclopedias. The Arabs had little interest in Greek literature but did translate much of Greek science and philosophy, including most of Aristotle. Among the Arabs he possessed an authority he never had in antiquity. It was really only with the twelfth and thirteenth centuries that this body of learning entered the West with translations from the Arabic and the Greek into Latin. The two greatest translators from the Greek were James of Venice (fl. 1136–1148) and William of Moerbeke (c. 1215–c. 1286), a Flemish Dominican who served as a chaplain and confessor to several popes and worked extensively in Greece becoming archbishop of Corinth there in 1278. James translated Aristotle's *Physics*, *On the Soul*, the *Parva Naturalia*, and part of the *Metaphysics*, and Moerbeke translated some forty-eight treatises (including seven on mathematics and mechanics by Archimedes), making available the remainder of Aristotle's natural philosophy. He also translated Greek commentaries from antiquity on Aristotle, such as Simplicius' commentary on Aristotle's *On the Heavens*. The greatest translators from Arabic were Gerard of Cremona (c. 1114–1187), who translated some seventy-one works, including most of Aristotle's natural philosophy, but also Ptolemy's *Almagest*, and twenty-four medical works, and Michael Scot (1175–c. 1232), who translated not only parts of Aristotle but some of the commentaries of the Spanish Muslim philosopher Averroes or Ibn Rushd (1126–1198) ("The Commentator") on Aristotle's works. These translations transformed the curriculum of the medieval universities, and by the thirteenth century Aristotle's logic and natural philosophy formed the core of that curriculum.

One of the great early figures in the growth of scholasticism was Peter Abelard (c. 1079–c. 1142), the most brilliant and the most arrogant teacher Paris ever saw, and who made Paris' reputation as an intellectual center in the first half of the twelfth century. He became a teacher at the cathedral school of Notre Dame about 1113, and soon students came from as far away as Italy, Germany, and England. He was a master of logic and debate, and his most famous book, *Sic et Non (Thus and Not)*, is a collection of 158 questions of philosophy and theology with contrary passages from the Fathers of the Church for each topic but no resolution. In the preface, however, he gives some general principles on how to proceed. "By doubting we come to examine, and by examining we reach the truth."

The triumph of Aristotle, of dialectic, and of speculative thought was of major significance, moving education from a more fundamentally humanistic, rhetorical, and literary mode as we saw in the classical world through the Carolingian revival of the eighth and ninth centuries, and the "Renaissance of the Twelfth Century" (to use the famous phrase of the historian Charles Homer Haskins), into something fundamentally philosophical and scientific. While we tend to

associate an interest in classical literature with the humanists of the
fourteenth century and later, in fact the schools of the eleventh and
twelfth centuries also strove to revive and make fuller use of the classical literary authors, particularly at the schools of Chartres and Or-
leans. John of Salisbury (c. 1120–1180), who began his studies in Paris
under Peter Abelard, was also well-read in many of the great classical
Latin authors (Cicero, Vergil, Lucan, Ovid, Horace, Juvenal, Persius,
Martial, Terence, Petronius, Sallust, Suetonius, Seneca, Quintilian,
Pliny the Elder) and wrote with a pure, flexible, and natural Latin
style. Saint Bernard of Clairvaux (1090–1153), the ascetic monk and
major cultural figure who saw the developing scholasticism as dangerously rationalistic and pursued Abelard until he was condemned
at the Council of Sens in 1140, had a command of classical literature
and mastery of Latin style that made him one of the powerful and
eloquent writers of his age.

Aristotle also affected theology, "the Queen of the sciences". Since
every student who studied theology had first to pass through the arts
curriculum with its Aristotelian logic and natural philosophy, it was
not surprising that Catholic theology, which had generally had an
Augustinian and Platonic emphasis before, moved into a new direction, with a greater appreciation of the physical and natural world,
a greater understanding of its autonomy, and a somewhat more
optimistic view of nature and reason. The founder of this Christian Aristotelianism was the German Dominican Albertus Magnus
(c. 1200–1280), Albert the Great—and he was called that even in
his lifetime. He was a theologian, philosopher, and natural scientist,
a *Doctor Universalis* ("the teacher of everything there is to know")
whose interest in the processes of nature led him to be considered
by some as a wizard and magician. Of the lesser German nobility, he
studied at Padua, where he mastered Aristotelian natural philosophy
and joined the Dominicans in 1223. He wrote commentaries or para-
phrases on all the Aristotelian books then available—more than eight
thousand pages in the nineteenth-century edition of his works.[17] This
was not to say that he was a pure Aristotelian, for he had a strong

---

[17] David C. Lindberg, *The Beginnings of Western Science: The European Scientific Tradition in Philosophical, Religious, and Institutional Context, 600 B.C. to A.D. 1450* (Chicago: University of Chicago Press, 1992), 229.

element of Neoplatonism in his thought, but he spread the influence of Aristotle within scholasticism and produced some important disciples, the most famous being Saint Thomas Aquinas (1225–1274), who also wrote many commentaries on Aristotle besides his great work of theology, the *Summa Theologica*. For these men, faith and reason each had its own integrity and were compatible ways to truth, though faith completed what reason could not reach and "grace perfects nature".

There was much in Aristotle that did not mesh with Christianity (e.g., the eternity of the world and the mortality of the human soul), and there were always those who saw in this acceptance of Aristotle a dangerous rationalism and naturalism. A decree in 1210 of the provincial synod of Sens, which included the diocese of Paris and its university, stated that no lectures were to be held in Paris, either publicly or privately, using Aristotle's books on natural philosophy or their commentaries. This decree did not affect other universities, and it did not hinder the use of Aristotle's logical works, ethics, and politics. This ban lasted until about 1240, and by 1245 Roger Bacon was already lecturing on these forbidden books. By 1255 Aristotle's natural philosophy had become the core of the arts curriculum. But difficulties remained, and criticism by the more conservative theologians toward the masters of the arts faculty intensified so much that in 1270 Bishop Étienne Tempier of Paris (d. 1279) condemned, under pain of excommunication for those who would defend or teach them, thirteen propositions drawn from Aristotle and Averroes. Finally, in 1277 Tempier condemned 219 propositions (179 philosophical and 40 theological) under pain of excommunication for those who taught them or even listened to them being taught unless they presented themselves to him or the university chancellor within seven days. A majority of them were drawn, in some way, from Aristotle's natural philosophy, of which the most significant were those dealing with the eternity of the world, the possibility of a double truth, and any limitation of God's absolute power—that is, his power to do anything he pleases short of a logical contradiction. Among those affected by some of these condemnations, though he had died some three years earlier, was Thomas Aquinas (some fifteen to twenty propositions), who sought a middle ground between the more radical arts masters who exalted Aristotle and the more conservative theologians who saw Aristotle as a threat. In 1325, nineteen months after Thomas Aquinas was officially declared a saint, those condemnations that touched upon his teaching were revoked, but the others remained in effect and were taken seriously through the fourteenth century.

The defense and promotion of God's absolute power (*potentia dei absoluta*) is of particular importance. While the general idea of the distinction between God's absolute power and God's ordained power (*potentia dei ordinata*) goes back to the Church Fathers, that pair of terms only first appeared together with Alexander of Hales in the twelfth century. Saint Thomas Aquinas used the distinction to maintain the freedom of God from any necessity. God's absolute will was what God could possibly have done short of contradiction, and his ordained will was what he did in fact do and would reliably continue to do. The distinction became a commonplace, but it took on a whole new significance and strength after the 1277 condemnations. On the one hand, it stimulated speculation by getting people to think beyond Aristotelian natural philosophy and its supposed impossibilities, to imagine a far broader range of hypothetical possibilities than would have been allowed before. This was certainly a plus. On the other hand, by emphasizing God's absolute freedom and power, his inscrutable will, his capacity to do whatever he wanted no matter how improbable or impossible short of a logical contradiction, it also encouraged a certain uncertainly about the capacity to find truth by natural means, by reason, logic, and science. This was particularly true, as it happened, as the range of what was considered a contradiction lessened and God's absolute power moved from a road not taken, as it had been before, to that of a presently active reality, ever-ready to override God's present ordained order, as it became.[18]

This tendency to set God over against his creation, of seeing the world's order not as some sort of participation in the divine reason or wisdom as we see in Saint Thomas Aquinas but as an arbitrary manifestation of the divine will, created a certain doubt about our ability to grasp reality with our minds and encouraged people to see faith and

---

[18] Francis Oakley, "The Absolute and Ordained Power of God in Sixteenth and Seventeenth Century Theology", *Journal of the History of Ideas* 59 (July 1998): 437–61; Francis Oakley, *Omnipotence and Promise: The Legacy of the Scholastic Distinction of Powers* (Toronto: Pontifical Institute of Medieval Studies, 2002).

revelation as the only means of finding certitude either about the world or about God. Even if God did not generally intervene at will, as most were usually quick to say, but followed the laws of nature of his own making, followed his own ordained power, it still created a certain sense of contingency and uncertainty, a lack of confidence in reason, that was very different from what we saw in the thirteenth century and very different from the great confidence we would see in the great figures of the Scientific Revolution, men like Copernicus, Galileo, and Newton, that the world's essential structure could be known.[19]

The idea of God's absolute power would also play an important part in the Galileo case. Pope Urban VIII demanded that Galileo put a strong version of the divine omnipotence argument into his *Dialogue*: that since one could not limit the power and wisdom of God, no argument about how the world works could ever be conclusive since God could always do something in the physical world inconceivable to the human intellect. The pope was highly attached to it, seeing it as irrefutable, but when Galileo showed a lack of respect for it by placing it toward the very end of the dialogue, in the mouth of the foolish pedant Simplicio, and with a rather weak response, it would become a major mark against him, and a key element leading to the trial of 1633.

In reference to the use of Scripture and natural philosophy, the scholastics generally followed the strictures of Saint Augustine, and they often spoke of the accommodation made to the limitations of the people of that time and to the common way of speaking. Thus, for example, despite what they read in the Bible, they knew that the earth was a sphere and described it as such. As the eminent historian of science Edward Grant has noted: "Virtually no theologians or natural philosophers from the thirteenth to fifteenth centuries interpreted biblical descriptions of cosmological or physical phenomena in a literal sense."[20] Even the famous passage of Joshua 10:12–14,

---

[19] Edward Grant, "Science and Theology in the Middle Ages", in *God and Nature: Historical Essays on the Encounter between Science and Religion*, ed. David Lindberg and Ronald Numbers (Berkeley: University of California Press, 1986), 57–59. William A. Wallace, "The Certitude of Science in Late Medieval and Renaissance Thought", *History of Philosophy Quarterly* 3, no. 3 (July 1986): 281–91.

[20] Edward Grant, *Science and Religion, 400 B.C. to A.D.: From Aristotle to Copernicus* (Westport, Conn.: Greenwood Press, 2004), 223.

a favorite of those opposed to Copernicus, where it states that God commanded the sun to stand still, could be seen in an accommodated sense. Nicole Oresme (c. 1320–1382)—philosopher, mathematician, and theologian at the University of Paris, bishop of Lisieux, adviser to the royal famly, and one of the most original of medieval scholastics—talking about this passage, could write in his *Livre du ciel et du monde* (1377):

> Also, when God performs a miracle we must assume and maintain that He does so without altering the common course of nature, in so far as possible. Therefore, if we can save appearances by taking for granted that God lengthened the day in Joshua's time by stopping the movement of the earth or merely that of the region here below—which is so very small and like a dot compared to the heavens—and by maintaining that nothing in the whole universe—and especially the huge heavenly bodies—except this little point was put off its ordinary course and regular schedule, then this would be a much more reasonable assumption.[21]

Despite the plain words of Scripture that God stopped the sun (and also against Aristotelian theory of a stationary earth and a rotating sphere of the fixed stars), Oresme argues that God would more likely act in the least disruptive and simplest manner possible, and that the stopping of a real daily axial motion of the earth would be more fitting. Such passages of Scripture, he wrote, could just be God accommodating himself to us, using the customary way of common speech just as "it is written that God repented, and He became angry and became pacified, and other such expressions which are not to be taken literally." In the end, however, not being able to demonstrate a moving earth, and following Saint Augustine's maxim that without demonstrative arguments one should follow what Scripture seems to say, he followed the traditional interpretation against the daily rotation of the earth.[22]

While many earlier scholars saw humanism as antithetical or at least as indifferent to science, and one cannot deny its primarily literary,

---

[21] Edward Grant, *A Source Book of Medieval Science* (Cambridge, Mass.: Harvard University Press, 1974), 509.

[22] Grant, "Science and Theology in the Middle Ages", 65–67.

rhetorical, and philological emphasis, still it influenced and even facilitated the birth of modern science in a number of ways.[23] It influenced it by its many translations and commentaries on classical texts, including many texts of philosophy and science—some for the first time—which increased their circulation and dissemination. We have already seen how it transformed Aristotelianism by making available the Greek texts of Aristotle, by its new translations of him, and by its translation of many of his ancient Greek commentators. It also changed the whole culture with a new attitude toward research in general so that even where older translations existed they were approached differently, with a new mature and critical historical sense, using philological skills to get to the original meaning of the texts in their proper historical context and in relation to other philosophical and scientific traditions of the time. Part of this new humanist attitude was its focus on practicality. The humanists had often criticized the impractical nature of much of the university curriculum, and this emphasis on usefulness had a positive effect in reference to science, both in approaching older scientific texts and in approaching the useful technology of Renaissance artisans. Archimedes, whose revival in the sixteenth century was so important to Galileo and so many others, had for the most part been long translated, much going back to the great wave of translation in the twelfth and thirteenth centuries, yet he had been little utilized. In the Middle Ages Archimedes was known primarily for how he died (so absorbed in a mathematical problem that he ended up being killed by a Roman soldier), but in the humanist retrieval of the sources he was portrayed as a brilliant creator of useful instruments and machines, and this led to a new interest and appreciation of his works.[24] This humanist focus on usefulness not only made humanism not opposed to a science based on reason and experience, but also positively receptive to it, particularly in reference to the technology which was being developed by Renaissance artisans. In Florence, both humanism and science developed productively

[23] Cesare Vasoli, "The Contribution of Humanism to the Birth of Modern Science", *Renaissance and Reformation* 3 (1979): 1–15; Anne Blair and Anthony Grafton, "Reassessing Humanism and Science", *Journal of the History of Ideas* 53 (1992): 535–40; Eric Cochrane, "Science and Humanism in the Italian Renaissance", *The American Historical Review* 81 (1976): 1039–57.

[24] W. R. Laird, "Archimedes among the Humanists", *Isis* 82 (1991): 628–38.

and even on occasion met in the same person, as in Brunelleschi, the creator of the great Dome of Florence's Duomo, and Leonardo da Vinci, with humanism supplying technology with a theoretical framework, elevating it to the rank of philosophy. Finally, humanism aided the development of science by its destruction of the argument from authority (with its historical and critical mind-set), and by its emphasis on rhetoric it forced science to address the widest possible audience, encouraging outsiders to participate in the endeavor and persuade them of its significance and usefulness.

The influence of humanism was also felt in astronomy. In 1451 George of Trebizond (1396–1486), one of the many Greek scholars who came to Italy in the fifteenth century and was involved in the great wave of new translations, translated Ptolemy's *Almagest* into Latin from the Greek. In 1460 Cardinal Bessarion (1403–1472), another famous Greek émigré scholar now in the service of the papacy as the legate of Pius II, arrived in Vienna, and considering Trebizond's translation of Ptolemy inferior, convinced Georg Peurbach (1423–1461) to undertake a Latin abridgment of Ptolemy's work. Peurbach taught classics and astronomy at the University of Vienna, had been court astrologer to King Ladislas V of Hungary and Bohemia, and was now court astrologer to Holy Roman Emperor Frederick III. He had already produced in manuscript his *Theoricae Novae Planetarum* (*New Theory of the Planets*) in 1454, which was a great improvement over the older work from the twelfth century of Gerard of Cremona, becoming the standard university text for astronomy. In 1459 he had produced the *Tabulae Eclipsium* (*Tables of Eclipses*), which listed lunar and solar eclipses for decades ahead, an invaluable tool for astronomers and astrologers, and which was widely read in manuscript before being first printed in 1514. Peurbach began this *Epitome* of the *Almagest* (he summarized the first six books of Ptolemy), but it was his brilliant student Johann Müller (1436–1476)—also known as Regiomontanus from the latinized place of his birth, Königsberg (King's Mountain) in Franconia—who completed it in 1462. The *Epitome* was not merely a compressed translation but included more recent observational data as well as corrections and critical commentary.

Regiomontanus became a member of Cardinal Bessarion's entourage (Bessarion possessed the largest private library then in Europe and one particularly rich in Greek mathematical manuscripts) and served him from 1461 to 1465, following him to Italy. It was in this circle that he learned Greek. He probably finished the *Epitome* in 1462, though it was only printed in 1496. In 1464 he completed a work on trigonometry, *De triangulis omnimodis* (*On Triangles of Every Kind*), and in 1467 went off to Hungary, building astronomical instruments and producing two influential works: the *Tabulae directionum* (*Table of Directions*), which helped to determine the positions of the planets based on the daily rotation of the night sky, and a table of sines. In 1471 he moved to Nuremberg, where the rich merchant Bernhard Walther financed not only the construction of one of the first formal observatories in Europe but also a printing press to publish mathematical and scientific works, the first being Peurbach's *Theoricae Novae Planetarum* in 1472. While he planned to publish the classics of Greek science in Latin translations and other scientific works, only a few were completed at his death, including his *Ephemerides* (1474), which showed the daily planetary positions from January 1, 1475, to December 31, 1504—which later came in handy for Columbus, who was able to overawe the Indians of Jamaica by predicting an eclipse. In Regiomontanus, we see a true fusion of humanist ideals with the tradition of university mathematics.[25]

Regiomontanus' work, his *Epitome* particularly, had an enormous influence on Nicolaus Copernicus (1473–1543), a young Polish cleric from a good burgher family in Toruń in Royal Prussia, part of the Kingdom of Poland since 1466. Copernicus first studied at the University of Kraków (mostly mathematics and astronomy) and then went to Italy in 1496 to study canon law at Bologna, where he learned Greek and much astronomy instead, living and working with the astronomy professor Domenico Maria Novara. It was with Novara that he made his first known celestial observation when at 11 P.M. they saw the eclipse by the moon of the star Aldebaran. Copernicus spent the Jubilee Year (1500) in Rome, where he gave public lectures on mathematics before briefly returning to Poland in 1501, without a degree. He returned soon afterward to study medicine at Padua, though he instead received a doctorate in canon law from Ferrara

[25] James Steven Byrne, "A Humanist History of Mathematics? Regiomontanus's Padua Oration in Context", *Journal of the History of Ideas* 67, no. 1 (January 2006): 41–61.

in 1503. He finally returned to Poland to serve his maternal uncle the bishop of Warmia, his patron, and between 1510 and 1514 he wrote his short *Commentariolus*, which first mentions his heliocentric theory of the universe, though it was never printed but sent out in manuscript to friends. In 1543 he published in six books his mature work on the subject, following the layout of the *Almagest*, which it updated, the *De revolutionibus orbium coelestium* (*On the Revolutions of the Heavenly Spheres*).

While Copernicus' system was an improvement on the Ptolemaic, being simpler and more economical, it still required eccentrics and epicycles, just fewer of them, reducing them from fifty-five to thirty-four. Having added no new observational data, he was unable to show his system to be superior much less certain, and so he was worried about its reception. In his own preface where he addresses Pope Paul III, to whom the work is dedicated, he writes that many would object to its novelty:

> I am aware that a philosopher's ideas are not subject to the judgement of ordinary persons, because it is his endeavor to seek the truth in all things, to the extent permitted to human reason by God. Yet I hold that completely erroneous views should be shunned. Those who know that the consensus of many centuries has sanctioned the conception that the earth remains at rest in the middle of the heaven as its center would, I reflected, regard it as an insane pronouncement if I made the opposite assertion that the earth moves. Therefore I debated with myself for a long time whether to publish the volume which I wrote to prove the earth's motion or rather to follow the example of the Pythagoreans and certain others, who used to transmit philosophy's secrets only to kinsmen and friends, not in writing but by word of mouth, as is shown by Lysis' letter to Hipparchus. And they did so, it seems to me, not, as some suppose, because they were in some way jealous about their teachings, which would be spread around; on the contrary, they wanted the very beautiful thoughts attained by great men of deep devotion not to be ridiculed by those who are reluctant to exert themselves vigorously in any literary pursuit unless it is lucrative; or if they are stimulated to the nonacquisitive study of philosophy by the exhortation and example of others, yet because of their dullness of mind they play the same part among philosophers as drones among bees. When I weighed these considerations, the scorn which I had reason to fear on account of the novelty and unconventionality

of my opinion almost induced me to abandon completely the work which I had undertaken.[26]

But in the end his friends and some other scholars convinced him to publish despite the unfounded and ignorant criticism he would receive. He particularly criticized those who would distort Scripture and use it against him, pointing particularly to the example of the Latin Church Father Lactantius (c. 250–c. 325), who spoke childishly when he mocked those who thought the earth spherical, since astronomy is written for astronomers.

Early in Galileo's *Letter to the Grand Duchess Christina*, written and circulated in 1615 but not published until 1636, he gives the early history of the relationship of Copernicus and the heliocentric theory with the Catholic Church. He tells us that Copernicus was not only a Catholic but a priest and a canon who was so highly regarded that he was called to Rome "from the remotest parts of Germany" when under Pope Leo X the Lateran Council was discussing the reform of the calendar; and that he was charged by the bishop of Fossombrone, then in charge of this undertaking, to try to master the science of celestial motions so as to make the proper corrections. By a herculean effort he did so and "then in accordance with his doctrine not only was the calendar regularized, but tables of all planetary motions were constructed." He published this at the request of the cardinal of Capua and the bishop of Kulm, and "since he had undertaken this task and these labors from the order of the Supreme Pontiff, he dedicated his book *On Heavenly Revolutions* to the successor of the latter, Paul III. Once printed this book was accepted by the Holy Church, and it was read and studied all over the world without anyone ever having had the least scruple about its doctrine."[27]

While there is much here that is true, there is also much that is erroneous.[28] Copernicus was a Catholic and a canon—of the cathedral chapter of Warmia at Frauenburg (Frombork), where his uncle (who had raised and educated him, and to whom he was secretary

[26] Nicholas Copernicus, *On the Revolutions*, ed. Jerzy Dobrzycki, translation and commentary by Edward Rosen (Baltimore: Johns Hopkins University Press, 1978), 7.

[27] Maurice A. Finocchiaro, *The Galileo Affair* (Berkeley: University of California Press, 1989), 89–90.

[28] Edward Rosen, "Galileo's Misstatements about Copernicus", *Isis* 49 (1958): 319–30.

and physician) was prince-bishop. He does not appear to be a priest, though he must have been in minor orders to have received an ecclesiastical benefice as a canon; and while his ancestors who came east into Poland in the thirteenth century may have been German, they had lived in Polish lands for a long time. While the Lateran Council (1512–1517) did discuss calendar reform (it would come up again at the last session of the Council of Trent) and in 1514 Pope Leo X did summon to Rome experts to help reform the calendar, Copernicus was not among them. Paul of Middleburg (1445–1533), the bishop of Frossombrone, was in charge of this, and he received many responses from scholars on how to proceed, including one from Copernicus, though it is now lost; and while the book was dedicated to Pope Paul III, no pope asked Copernicus to undertake it. The cardinal of Capua and the bishop of Kulm did encourage him to publish. Nicholas von Schönberg (1472–1537), made archbishop of Capua by Pope Leo X in 1520 and a cardinal under Paul III in 1535, did write a highly laudatory letter in 1536 to Copernicus, asking him to share his new astronomy and cosmology with others and to send him his research so that he could have a copy. And the bishop of Kulm (Chelmno), Tiedemann Giese (1480–1550), a close friend of Copernicus and buried next to him at Frauenberg Cathedral, was constantly trying to get Copernicus to publish his work.

It is not correct, however, to say as Galileo and others have said that the Gregorian calendar was regularized in accordance with Copernicus' doctrine. While the Gregorian reform commission adopted Copernicus' measurement of the length of the year and his theory of precession of the equinoxes, the heliocentric theory was not involved. The Gregorian calendar reformers deliberately did not connect the calendar to any particular astronomical hypotheses, and Copernican cosmology was in no way essential to the new calendar. If anything, as Edward Rosen has pointed out, the Gregorian calendar was regulated in conformity with the doctrine of Luigi Giglio (Aloisius Lilius), whose plans Pope Gregory XIII explicitly mentioned in the bull announcing the new calendar.[29]

---

[29] James M. Lattis, *Between Copernicus and Galileo: Christoph Clavius and the Collapse of Ptolemaic Cosmology* (Chicago: University of Chicago Press, 1994), 3–4 and n. 3, 221–22; Rosen, "Galileo's Misstatements about Copernicus", 328–29.

Finally, despite what Galileo says, the book was not accepted by the Church and studied all over the world without the least scruple by any one. While Paul III, to whom he dedicated his book, certainly showed a great interest in the stars, as we know from his extensive patronage of the renowned astrologer Luca Gaurico (1476–1558), who had predicted his election and whom he made a bishop, he seems not to have shown any interest in Copernicus.[30] Certainly, the years after its publication were very busy ones for Paul III as he tried to convene the Council of Trent, so it is not surprising if he took no notice of the book. Many were also fooled by the anonymous *Letter to the Reader* by Andreas Osiander (1498–1552), who had seen to the printing of the book in Nuremberg. This prefaced Copernicus' work and was added without Copernicus' permission and against his wishes. It stated that the heliocentric theory should be seen as a convenient fiction and not as describing reality. Anyone who read the book more seriously would have seen that that was not what Copernicus was saying. In fact, it is relatively clear that Copernicus believed that he was describing reality.

Among those who took notice of the work and saw that it was meant to be taken as describing the real world, and were not pleased in that, were the Florentine Dominican theologian Giovanni Maria Tolosani (1470/1471–1549) and his close friend, another Dominican, Bartolomeo Spina (c. 1475–1547). Born in Pisa and joining the Dominicans in 1493, Spina had taught theology within his order and at the University of Padua, as well as serving briefly as inquisitorial vicar in Modena (1517–1519), where he prosecuted witches. Besides editing many of Saint Thomas Aquinas' works (some of his commentaries on Aristotle and his biblical commentaries), he wrote against the dangers of witchcraft and attacked the views of the soul of the Aristotelian philosopher Pietro Pomponazzi and of his own Dominican confrere Cardinal Cajetan. Appointed in 1545 (not 1542 as some have it) by Paul III as the Master of the Sacred Palace, the pope's theologian in the Roman Curia, he was so esteemed by Paul III that he was in the inner circle of his advisors over matters touching on the Council of Trent. Tolosani wrote that Spina had planned to have

[30] Robert S. Westman, *The Copernican Question: Prognostication, Skepticism, and Celestial Order* (Berkeley: University of California Press, 2011), 134.

the book condemned, but illness and death intervened. Tolosani, himself an accomplished astronomer who had written on calendar reform, wrote a detailed criticism of Copernicus, philosophically and theologically, showing him to be dangerous to the faith. But this work was never published and remained in manuscript, though it would be read later by the Dominican Tommaso Caccini when Galileo started to promote the heliocentric theory.[31]

That being said, there were also those favorable to Copernicus, such as Johann Albrecht Widmannstetter (1506–1557), German humanist, orientalist, and biblical scholar, and secretary to Pope Clement VII and later to Pope Paul III. He outlined the Copernican theory in the Vatican Gardens before an intimate gathering of Pope Clement VII, two cardinals, a bishop, and the pope's physician in the summer of 1533. Pope Clement VIII was so pleased by it that he rewarded Widmannstetter with a valuable Greek codex containing several philosophical treatises. In 1535 Widmannstetter would become secretary to Cardinal Nicholas von Schönberg, who, as we have seen, wrote encouragingly to Copernicus.

While many used Copernicus' calculations for their greater accuracy concerning the planetary movements—particularly after Erasmus Reinhold (1511–1553) put together and published in 1551 the Prutenic Tables, astronomical tables based on Copernicus' calculations—the heliocentric theory was generally absolutely rejected as physically false and absurd. Copernicus' work was held in high esteem at the University of Salamanca, where as early as the 1561 statutes of the university, the chair of mathematics and astrology allowed for the reading of Copernicus in the second year if the students voted for it. In the 1594 statutes the teaching of Copernicus was reaffirmed, even without a vote of the students, though the way it was phrased it suggested that it was the technical aspects that were utilized, as was true at a number of universities, and not the heliocentric aspects. That the 1625 statutes repeat the text of the 1594 statutes supports the idea that it was on the technical

[31] Michel-Pierre Lerner, "The Heliocentric 'Heresy' From Suspicion to Condemnation", in The Church and Galileo, ed. Ernan McMullin (Notre Dame, Ind.: Notre Dame University Press, 2005), 14–16; Westman, Copernican Question, 195–97; Edward Rosen, "Was Copernicus' Revolutions Approved by the Pope?" Journal of the History of Ideas 36 (1975): 531–42.

aspects that he was considered since the 1616 judgment of the Roman Inquisition against Copernicanism had already come out.[32] The only positive theological response to Copernicanism in the Catholic world, at least before the Italian Carmelite theologian Paolo Antonio Foscarini's letter of 1615, was that of the Spanish Augustinian Diego de Zúñiga (1536–1598) and the Italian Jesuit Benedetto Giustiniani (c. 1550–1622). Zuñiga, who possessed a copy of *De revolutionibus* and had a competent knowledge of astronomy, was educated at Salamanca and Alcala, and from 1573 until about 1580 taught Scripture at the University of Osuna. In his *Commentaries on Job* (1584), looking at the passage at Job 9:6, "He who moves the earth from its place, and its pillars are shaken," he wrote that this passage was better explained by the heliocentric system, and that in general the Copernican system was better than the Ptolemaic on scientific grounds. He gave technical reasons for all this and wrote that the passages commonly used to refer to the motion of the sun in Scripture were just the common way of speaking and should not be taken as scientific. In 1592 Giustiniani, commenting on the astronomical positions of the philosopher Francesco Patrizi, and clearly influenced by Zuñiga's commentary, believed that Copernicanism could "without difficulty" be accommodated to Scripture and that it was really only an issue between philosophers. Both, however, would change their minds. In his last published work, *Philosophia prima pars* (1597), the only published part of what would have been a three-part discussion of the whole of philosophy, Zuñiga subjected the heliocentric theory to a careful philosophical and scientific critique and came to the conclusion that it was absurd and impossible. In 1616 Giustiniani would be among the Inquisition's theological consultants who would condemn the two Copernican propositions that the immovable sun is at the center of the world and that the earth is not at the center of the world and does move.[33]

Not long after the publication of Zúñiga's *Philosophia*, the Jesuit Juan de Pineda (1558–1637), in his own commentary on Job

[32] Victor Navarro Brotóns, "The Reception of Copernicus in Sixteenth-Century Spain: The Case of Diego de Zúñiga", *Isis* 86 (1995): 55–60.

[33] Victor Navarro Brotóns, "The Reception of Copernicus in Sixteenth-Century Spain: The Case of Diego de Zúñiga", *Isis* 86 (1995): 52–78; and Lerner, "The Heliocentric 'Heresy'", 17–18.

(1598–1602), criticized Zúñiga's Copernican interpretation: "We will not say anything further now concerning this opinion except that it is plainly false (others indeed say that it is foolish, frivolous, reckless, and dangerous to the faith and that Copernicus and Caelio Calcagnini have revived it from the dead remains of those ancient philosophers [Pythagoreans] more as a figment of their own imaginations than as something good and useful for philosophy)."[34] Other Jesuits, such as Jean Lorin (1559–1634) and Nicolas Serarius (1555–1609), also condemned Copernicus. But the most significant Jesuit to have opposed Copernicanism was the order's greatest mathematician and astronomer: Christopher Clavius.[35]

Clavius was the last great expositor of the Ptolemaic system, and in his *Sphaera*, which through its many editions eventually became quite a substantial book, he not only defended and explained the virtues of that model but dealt with objections against it. He also discussed and criticized in some detail other, rival cosmologies. We usually think only of two systems existing at this time, the dominant Ptolemaic and the new-fangled Copernican, with the Tychonic coming later and splitting the difference—and the standard works on astronomy generally imply the same. Galileo in his *Dialogues on the Two Great World Systems* strengthened this perception that there were only two theories out there, but Galileo had strong reasons to ignore all other systems but the Copernican and Ptolemaic since it suited his polemical aims to leave Copernicanism with the mastery of the field. In fact, there was quite a variety of cosmological theories that struggled for dominance in the sixteenth century and which Clavius discussed and criticized.

The first of these were the homocentric theories where each of the planetary spheres was homocentric with the earth; that is, they each had the earth as their same center, unlike the epicyclic and eccentric models with multiple centers as we see in Ptolemy. These theories began with Eudoxus and Callippus in the fourth century B.C. (they were therefore contemporaries of Plato and Aristotle) and predate the models of Hipparchus in the second century B.C., which

---

[34] Richard J. Blackwell, *Galileo, Bellarmine, and the Bible* (Notre Dame, Ind.: Notre Dame University Press, 1991), 26–27.
[35] Lerner, "The Heliocentric 'Heresy'", 18–19.

used eccentrics and epicycles. The homocentric model was also the one taken up by Aristotle. It was also taken up later by Averroes, some medievals, and in the early sixteenth century by a number of thinkers associated with the University of Padua such as G. B. Amico and Girolamo Fracastoro, with Fracastoro being the most significant. Girolamo Fracastoro (1483–1553) was from an ancient family of Verona in the Republic of Venice and was a famous physician—Pope Paul III appointed him as the official physician to the Council of Trent. He studied at Padua, where his teachers included the Aristotelian philosopher Pietro Pomponazzi, and became a professor there at nineteen. Besides writing extensively on philosophical and medical topics (the term "syphilis" comes from him), he also published in 1538 his *Homocentrica* (dedicated to Paul III), proposing a homocentric view of the cosmos. These theories always remained very much of a minority position since they, in short, didn't work. They failed to explain adequately what astronomers saw in their observations of the planetary orbits, and they failed to have any real predicative value—you could not accurately ascertain where the planets would be in the future.[36]

The second rival cosmology was the fluid-heaven theory: that the planets were not rigidly fixed within celestial spheres, or that there were no celestial spheres at all, but that the planets were in concentric zones of a fluid medium through which the planets moved on their own, moving through the heavens like fish in water or birds in the air. While the Stoics believed that the heavens were filled with an animate fluid called *pneuma*, and the planets, being intelligent, directed themselves through this fluid *pneuma* (and, as we have seen, the revival of Stoicism in the Renaissance may have influenced the spread of fluid-heavens ideas), the first we see of fluid-heaven theories in the West is from the late thirteenth century. The most important expositors of these in the sixteenth century were the Platonic philosopher Francesco Patrizi (1529–1597) and Clavius' friend and colleague at the Roman College, the great Jesuit theologian Robert Bellarmine (1542–1621).

As we have already seen, Patrizi occupied chairs of Platonic philosophy at Ferrara and Rome (called by Pope Clement VIII), and

---

[36] Lattis, *Between Copernicus and Galileo*, 87–94.

attempted to replace Aristotle, but his great work *Nova de univer-sis philosophia* was denounced, and the Congregation of the Index prohibited it until corrected, with all available copies destroyed. It had many anti-Aristotelian elements including replacing the four ele-ments with another four of space, light, heat, and fluid or humid-ity and abolishing the solid spheres to let the planets move freely in ether, "fly within a liquid sky".[37]

When Bellarmine was at Louvain in the Low Countries from 1570 to 1576 and lecturing on Saint Thomas Aquinas' *Summa Theologica* (as was the Jesuit way), he discussed the questions in the *Summa* on the six days of creation (Hexameron) and put together a cosmology that was profoundly biblical and anti-Aristotelian, and profoundly skep-tical of late Ptolemaic cosmology. While he never published these lectures, he seems never to have rejected them as a 1618 letter to Prince Cesi reveals.[38] Denying Aristotle's immutability of the heav-ens and the distinction between the terrestrial and celestial realms, Bellarmine kept a geocentric and geostationary cosmos with a single solid sphere bearing the fixed stars and a fluid substance filling all the space between the fixed stars and the earth, the planets moving freely and by themselves through this fluid "like birds in air or fish in the sea". Clavius, who never mentions Bellarmine by name in any of his criticisms of the fluid-heavens cosmology, denied that planets move by themselves but according to the orb in which they are fixed "like a knot in a board moves with the board or a nail in a wheel moves with the wheel" and criticized the fluid-heavens theories even more than the homocentric ones since if they moved like fish or birds there was no way you could predict what the motions of the planets were and therefore where they would be in the future, lack-ing even the grounding in mathematical concepts that we see in the uniform circular motion of homocentric theories.[39]

[37] Paul Oskar Kristeller, *Eight Philosophers of the Italian Renaissance* (London: Chatto & Win-dus, 1965), 110–26.

[38] Blackwell, *Galileo, Bellarmine, and the Bible*, 48–49.

[39] G. V. Coyne and U. Baldini, "The Young Bellarmine's Thoughts on World Systems", in *The Galileo Affair: A Meeting of Faith and Science*, ed. G. V. Coyne, M. Heller, and J. Zycinski (Vatican City, 1985), 103–10 (also in *The Louvain Lectures of Bellarmine and the Autograph Copy of His 1616 Declaration to Galileo* [Vatican City: Vatican Observatory, 1984]); and Lattis, *Between Copernicus and Galileo*, 96–102.

The third cosmology criticized by Clavius was associated with the fluid-heavens though it was superior to it since the planets did not move like fish in the sea or birds in the air but through some sort of channels, like the veins of an animal. He mentioned no names and considered that without any necessity or even probability it would be absurd to think "that the body of the heavens is perforated by such channels and filled everywhere by that fluid substance that heretofore no philosopher seems to have admitted".[40]

Finally, there was his criticism of Copernicus. Unlike the other cosmologies, it was, despite the fact that it had first been proposed by Aristarchus in the fourth century B.C., something new, and there was very little previous criticism to utilize. In his first edition of his *Sphaera*, in 1570, he refuted Copernicus briefly, but in the next edition of 1581 and thereafter, there was a more extended critique based on physical, methodological, and scriptural reasons. He considered that the Copernican position held many errors and absurdities such as saying that the earth was not in the center of the universe, that it moved by a triple motion, and that the sun stood in the center of the world, having no motion at all. All of this was contrary to the common teaching of philosophers and astronomers and opposed Sacred Scripture. That being said, there was also considerable praise of Copernicus: admitting that the Copernican system was as good as the Ptolemaic in predicting and explaining phenomena and that his mathematical calculations were of great value. Clavius was like Copernicus in that he was a realist, believing that heavenly phenomena could reveal real knowledge about the heavenly bodies, and while he criticized Copernicus, he gave the greater space to criticizing Fracastoro's homocentrics, which seems to have been for him far more dangerous, especially at a time when philosophical skepticism was strong and growing. As Clavius' most significant recent expositor put it: "To retreat to the philosophically seductive homocentric astronomy, which is fundamentally incapable of saving phenomena, is to abandon the search for physical explanations of the real celestial mechanisms and thus to abandon also the hope of true human comprehension of the created world. The triumph of skepticism, even in astronomy, could

[40] Lattis, *Between Copernicus and Galileo*, 102–4.

only be a defeat for an orthodoxy that claimed to know how the heavens go."[41]

Clavius' task of defending the Ptolemaic system became more difficult with each passing decade. A key event in the decline of the Ptolemaic world, and in the history of astronomy itself, was the supernova of 1572: a massive star exploding with a brightness millions of times greater than our sun, which was visible for about sixteen months in the constellation of Cassiopeia. It was this prodigy of a new star that brought the Danish aristocrat Tycho Brahe (1546–1601) to the fore as a world-famous astronomer. He early on realized its significance, observed and measured it carefully, and then published in 1573 at Copenhagen his De stella nova, which questioned one of the foundations of Aristotelian cosmology, showing that the heavens were not immutable since this star was clearly beyond the moon where any such change was judged impossible. Brahe, supported by King Frederik II of Denmark, then built the largest, most technologically sophisticated observatory in Europe at the island of Hven, first at Uraniborg in 1576 and then at Stajaerneborg in 1581, where he took the most precise measurements of the stars available in these pre-telescopic days and twice as accurate as his predecessors. In 1577 a comet appeared which was also beyond the moon, with the path of the comet showing the falsity of the idea of hard celestial spheres. In 1588, in his Astronomiæ Instauratæ Progymnasmata (Introduction to the New Astronomy), he published his own cosmological system, the geo-heliocentric, where all planets orbit the sun, and the sun orbits a perfectly still earth and its moon. Because he could find no stellar parallax (the apparent shift of position of any nearby star against the background of distant objects)—which he would have seen if earth was moving around the sun—he believed the earth must be immobile at the center of the universe, as Scripture also taught. With the death of Tycho's patron King Frederik II, he fell out of favor and left Hven in 1597 with all of his vast astronomical data, settling in Prague to be Imperial mathematician to Emperor Rudolph II (1552–1612). It was there that he connected with Kepler, who was to complete his work.

Tycho's geo-heliocentric system answered the observed astronomical difficulties, kept the deeply intuitive and commonsense

[41] Ibid., 144.

view that the earth was the center of the universe and did not move, and, very importantly, it did not conflict with Scripture. Even before the condemnation of Copernicanism in 1616, and even before Galileo's telescopic discoveries of 1610, it was starting to find adherents among the Jesuit scientists, and would later become their favored theory. Clavius, while praising Tycho's observational work, completely ignored it in his revisions of his *Sphaera* commentary, perhaps because it would mean he would have to abandon the solid celestial spheres and accept some sort of fluid-heavens cosmology, which he had opposed his whole life.[42]

Johannes Kepler (1571–1630) became the first great promoter of Copernicus among astronomers. Educated at Tübingen and trained by the astronomer Michael Maestlin (1550–1635), one of the first astronomers to subscribe to Copernicus' heliocentric theory (though in his regular undergraduate lectures he taught the Ptolemaic system), he became a Protestant schoolmaster in Graz. He worked with Tycho near Prague in 1600 and inherited his data and position as Imperial mathematician at Tycho's death. A Copernican from his youth, his 1596 *Mysterium Cosmographicum* (*Cosmographic Mystery*) was the first published work to defend the Copernican system. Kepler, however, was not the clearest of writers, and his science was mixed in with esoteric religious speculations. As Owen Gingerich wrote of the book: "Seldom in history has so wrong a book been so seminal in directing the future course of science."[43] In 1606 he published his *Stella Nova* (*The New Star*) on the new star that appeared in 1604, and in 1609 he published his *Astronomia Nova* (*A New Astronomy*), which contained the first two laws of planetary motion. The first law was that planets move in elliptical orbits, and the second was that a planet will sweep out equal areas in equal intervals of time. He argued that no theory which insisted on circular motion could account for the orbit of Mars, which was an ellipse. Galileo possessed a copy of the *Astronomia Nova*, but he seems not to have read it, or read it carefully, for there was much there which would have supported his case for Copernicanism. In his most important

---

[42] Ibid., 205–16.
[43] Owen Gingerich, "Johannes Kepler", *Dictionary of Scientific Biography*, ed. Charles Coulston Gillispie (New York: Charles Scribner's Sons, 1973), 7:289–312.

work, *Epitome astronomiae Copernicanae* (*Epitome of Copernican Astronomy*), published in three installments of 1617, 1620, and 1621, Tycho elaborated all three laws of planetary motion (the third law being that the square of the orbital period of a planet is proportional to the cube of the semimajor axis of its orbit) and attempted to explain the heavenly motion by physical causes.

To give you a sense of the times in which Kepler was living, in the very period when he was publishing his *Epitome*, his mother was charged as a witch, imprisoned for fourteen months, and finally, with considerable help from Kepler, freed in 1621. His mother's aunt, who had raised her, had been burned at the stake as a witch, and in that little town where this all took place, Leonberg in Baden-Württemberg, between 1613 and 1629 some eight women had been executed for witchcraft. Being different could be a dangerous thing in the early seventeenth century, and having powerful connections, in this case a son who was also Imperial mathematician, was probably the only thing that kept her from the stake.

In 1597 Galileo received a copy of Kepler's *Mysterium Cosmographicum*, and he wrote to Kepler on August 4, 1597, thanking him, stating that while he had read only as far as the preface, he was looking forward to the rest. He also stated that he had been a Copernican for some years but had not admitted it publicly for fear of the ridicule of colleagues. "Like you, I accepted the Copernican positions several years ago and discovered from thence the causes of many natural effects which are doubtless inexplicable by the current theories. I have written up many reasons and refutations on the subject, but I have not dared until now to bring them into the open, being warned by the fortunes of Copernicus himself, our master, who procured for himself immortal fame among a few but stepped down among the great crowd (for this is how foolish people are numbered), only to be derided and dishonored. I would dare publish my thoughts if there were many like you; but since there are not, I shall forbear."[44] Kepler replied and urged him publicly to acknowledge his Copernicanism, but Galileo did not respond, and there was no more contact between the two for thirteen years. The first sign of Galileo's support

---

[44] Giorgio de Santillana, *The Crime of Galileo* (Chicago: University of Chicago Press, 1955), 11.

for Copernicanism was slightly earlier than his letter to Kepler, in his letter of May 30, 1597, to Jacopo Mazzoni, professor of philosophy at the University of Pisa, where he wrote that he thought Copernicus' opinion considerably more probable than Aristotle's.

In the autumn of 1604 a new star appeared, and this nova aroused as much interest as that of 1572, which had inspired Tycho Brahe. Galileo gave three public lectures on the nova of 1604, which attracted a large audience. He proved that the nova was indeed far away, above the moon, and if there was change in the heavens then Aristotle and traditional astronomy were wrong. But it was not until Galileo heard about and made himself a telescope that his great astronomical career would take off. Elements of the telescope existed before its "invention" in Holland in 1608 (e.g., concave and convex spectacle lenses), and the Italians Giovanni Baptista della Porta (1535–1615), the great Renaissance scientist and occult philosopher, and Raffael Gualterotti (1548–1639) seem to have come upon something like the telescope but without any great magnification before 1600. But it was only in the Netherlands that through the work of someone, perhaps Hans Lipperhey, who first applied for the patent, perhaps another, since a number of people seem to have been independently working on this, a spyglass that could see at a significant distance was born. The technology was not complex and it could be easily copied, so it is not surprising that Galileo was able to build his own telescope. What is surprising is that Galileo was able to build telescopes of such great magnification so quickly and was able to adapt them for astronomical use. His first telescope magnified three times as much (nine-powered) as the original, and with this he wowed the Venetian Senate, winning an income increase and a lifetime appointment, but he soon did better—creating one by 1610 that magnified twenty times, one by the beginning of 1613 that magnified thirty times, and one by 1615 that magnified a hundred times. With his telescope now turned to the stars, he made his career and published his *Siderius nuncius* in 1610, which, to use the words of one scholar, "shook the world of learning and changed the course of astronomy."[45]

[45] Albert Van Helden, "The Invention of the Telescope", *Transactions of the American Philosophical Society* 76, no. 4 (1977): 27; and David Wootton, *Galileo: Watcher of the Skies* (New Haven, Conn.: Yale University Press, 2010), 92.

April 18, 1506, was an important date in Renaissance history. On that date was laid the foundation stone of the building that symbolized the Renaissance artistically and religiously: the new Basilica of St. Peter's in Rome. After twelve hundred years the old basilica was in such a state of severe disrepair that it was decided by Pope Julius II, despite great opposition, that a new building would arise to replace the old. The pope, accompanied by his cardinals and many prelates, went in procession behind the cross to the edge of the excavation, and entering the twenty-five-foot-deep hole, the pope blessed an inscribed foundation stone of white marble and set it with his own hands. Some masons then placed an earthen pot containing twelve special medals prepared for the occasion and stamped on one side with the head of Pope Julius and on the other with a representation of Bramante's new church.[46] While the day itself was chosen for the reason that it was the feast day of the founding of the original Basilica of St. Peter's by the emperor Constantine and Pope Saint Sylvester, the time of the ceremony, 10 A.M., was chosen because of the alignment of the stars: not only was it a most favorable and auspicious alignment astrologically, but it also had zodiacal connections to the birth of the world, the birth of Pope Julius II, and the birth of Jesus Christ.[47]

While it seems strange to us that the refounding of the premier church in Christendom should conform to astrology, it was seen as quite natural for contemporaries. We see astrology, the sister science to astronomy, as something superstitious or flaky, but for a great many centuries it was a highly respected science; as the venerable historian of science Lynn Thorndike has pointed out, before Newton's promulgation of the universal law of gravitation, there was the recognized and accepted universal natural law of astrology: that the entire world of nature was directed and governed by the movement of the heavens and the celestial bodies.[48] The belief in the immutable laws of nature and the first application of the mathematical method to natural phenomena were here—not in the Scientific Revolution of

[46] Ludwig Pastor, *The History of the Popes from the Close of the Middle Ages*, ed. Frederick Ignatius Antrobus, 2nd ed. (St. Louis: B. Herder, 1907), 473–74.

[47] Mary Quinlan-McGrath, "The Foundation Horoscope(s) of St. Peter's Basilica, Rome: Choosing a Time, Changing the *Storia*", *Isis* 92 (2001): 716–41.

[48] Lynn Thorndike, "The True Place of Astrology in the History of Science", *Isis* 46 (1955): 273–78.

the seventeenth century. In 1464 Regiomontanus, the greatest mathematician of his day, praised astrology as the queen of mathematical disciplines in his inaugural oration for his course of mathematical astronomy at Padua.[49] To a remarkable extent, the interest in astronomy and the ever-increasing accuracy of astronomical observation and discussion of the early modern period were ordered to astrological ends. That was certainly true for Tycho and Kepler, as well as many before them; and it is the thesis of Robert Westman's *The Copernican Question* (2011) that Copernicus' own work was an unstated attempt to strengthen astrology by correcting its astronomical/mathematical foundations, especially by a readjustment of the planetary order.[50] Galileo himself, as we will see, was a master astrologer.

The oldest extant individual horoscopes go back to the Babylonians in Mesopotamia of the fifth century B.C. As we have already seen, Ptolemy not only supplied medieval and Renaissance Europe with its astronomy, he also supplied it with its astrology—scientific astrology, in contrast to the amalgam of magic, demonology, and star worship that could also go under that name. In the literature of antiquity the terms "astrology" and "astronomy" are used interchangeably with "astrology" becoming the generic term for both the study of the movement of the stars and the study of the effects of these celestial movements on the earth. This semantic confusion would continue through the Middle Ages and into the Renaissance. In the Christian writer Isidore of Seville (c. 560–636) a distinction was made between natural astrology which included astronomical knowledge and astrological weather predictions and superstitious astrology which dealt with nativities (charts of the stars at one's birth) and predictions of one's future. It was particularly to this latter kind of astrology that the Fathers of the Church were very much opposed, and no one more than Saint Augustine, who wrote extensively against it, particularly in his *City of God*, though he devoted no single treatise to it. The ancients believed that the stars influenced our actions, and Aristotle and Plato as much as anyone, and in the

---

[49] H. Darrel Rutkin, "Astrology", in *The Cambridge History of Science*, vol. 3, *Early Modern Science*, ed. David C. Lindberg, Mary Jo Nye, Katharine Park, and Roy Porter (Historiker) (Cambridge: Cambridge University Press, 2008), 545.

[50] Westman, *Copernican Question*.

revival of ancient learning in the Middle Ages, the medievals followed them in this as in other areas.

One of the greatest defenders of a Christian form of astrology in the Middle Ages was Albertus Magnus.[51] While his astrological views are in almost all of his scientific works, his *Speculum astronomiae* (*The Mirror of Astronomy*), written between 1260 and 1270, is his most significant work on the subject. He defended astrology as a genuinely Christian form of knowledge. He believed that celestial influence began with God, his power then passing downward through the various spheres until it reached us, taking on accretions and imperfections along the way, and influencing us in a bodily way, in our physical impulses, and only indirectly affecting the soul, thus maintaining free will. Since the motions of the celestial bodies were uniform and predictable, successful astrological predictions were possible. Knowledge of astrology, therefore, could be quite useful in dealing with these influences and would, rather, be a good use of our free will, a perfecting of our free will, instead of a negating of it. Albertus Magnus liked to quote the dictum of the Persian astrologer Albumasar (787–886), though he erroneously attributed the concept to Ptolemy, that "the wise man will dominate the stars." He accepted most forms of predicative astrology, including revolutions, which were concerned with large-scale effects such as the weather or state affairs; nativities, which looked at people's future prospects based on the stars at their birth; interrogations, where answers to particular questions were sought; and elections, which were to guide people to choose the most propitious times to begin or do something.

Saint Thomas Aquinas followed Albertus Magnus in this. While he believed that using the stars to foretell the future was superstitious and wrong, and opened oneself to diabolic powers, and that man was always spiritually free to make his choices, astrology could be used to forecast future events that were really determined by physical laws, such as drought and rainfall, and celestial forces could dispose and

---

[51] See Scott Hendrix, *How Albert the Great's Speculum Astronomiae Was Interpreted and Used by Four Centuries of Readers: A Study in Late Medieval Medicine, Astronomy, and Astrology* (Lewiston, N.Y.: Edwin Mellen Press, 2010); and Paola Zambelli, *The Speculum Astronomiae and Its Enigma: Astrology, Theology and Science in Albertus Magnus and His Contemporaries* (Boston: Kluwer Academic Publishers, 1992).

incline man in his actions, making a bodily impression as it were, while leaving the soul free.[52]

Among the propositions condemned by Bishop Tempier of Paris in 1277, there were a number touching upon astrology. There were other academics, such as Nicole Oresme, who wrote in both French and Latin against it using both the standard arguments and also the more novel one of the impossibility of having accurate enough knowledge of the heavenly movements for correct predictions. Aware of his own king's addiction to astrology and divination, he says at the beginning of his *Livre de divinacions* (c. 1366):

> It is my aim, with God's help, to show in this little book, from experience, from human reason, and from authority, that it is foolish, wicked and dangerous even in this life, to set one's mind to know or search out hidden matters or the hazards and fortunes of the future, whether by astrology, geomancy [divination by interpreting markings on the ground or tossed handfuls of soil, rocks, or sand] and, nigromancy [black magic or black divination], or any other such arts, if they can correctly be called arts; and, further, that such things are most dangerous to those of high estate, such as princes and lords to whom appertains the government of the commonwealth.[53]

Astrology's greatest defender in the late Middle Ages, going further than his predecessors in using it to interpret history and prophecy, was Pierre Cardinal d'Ailly (1351–1420), a major theologian and philosopher at the University of Paris, a Church reformer and a leading figure in ending the Great Schism, which had been going on since 1378. He had begun as dismissive of astrology as of any effort of human reasoning, condemning most uses of judicial astrology, but from 1410 he wrote at least sixteen treatises dealing directly or indirectly with it, defending it with increasing openness and enthusiasm, praising its usefulness, and showing its compatibility with Christian theology and with human freedom.[54] He particularly focused on the great conjunctions of Saturn and Jupiter which occurred every 240

---

[52] *Summa Theologiae* Ia, q. 115, a. 4, and IIa–IIae, q. 95, a. 5; *Summa Contra Gentiles* 3, 82, 104.

[53] Grant, *Source Book of Medieval Science*, 488.

[54] Laura Ackerman Smoller, *History, Prophecy, and the Stars: The Christian Astrology of Pierre d'Ailly, 1350–1420* (Princeton, N.J.: Princeton University Press, 1994), 52.

and every 940 years. He saw astrology as a kind of natural theology, as an objective and certain science in understanding the divine plan and in interpreting Scripture and history, and particularly in predicting the time of the arrival of the antichrist. (It would occur in 1789.) Following many others, including Aquinas, he believed that the influence of the stars was not inconsistent with free will and divine omnipotence, affecting physical things, inclining but not necessitating the will. He even allowed for the influence of the stars on the human aspects of Christ's life and, therefore, the possibility of doing a horoscope for him.

Astrology was a common part of the science of the day, taught within the three disciplines of mathematics, natural philosophy, and medicine in the late medieval and Renaissance universities. The professor of astrology at the University of Bologna in the late fifteenth century, for example, was expected not only to teach his courses but also to participate in at least three public astrological disputations a year and to supply an annual prognostication (and other astrological services) for the university community.[55] Astrology was especially useful for medical students since every part of the body, every humor, every disease, and every drug was under the influence of the planets and ruled by the twelve signs of the zodiac, so that astrological knowledge assisted in the diagnosis of the disorder, the ascertaining of the cure, and setting the appropriate time for the treatment. Astrological almanacs were extremely popular: books with astronomical data of every sort in them as well as predictions about what would follow in the year, from weather conditions and public calamities to when the best time would be for planting crops or getting married, much like our present-day Old Farmers' Almanac.

The mathematics chair at Pisa included astrology, and lectures by the mathematics professor on the Tetrabiblios were given quite often. Galileo's predecessors at Pisa were very heavily involved in astrology. The Camaldolese monk Filippo Fantoni (c. 1530–1591), Galileo's immediate predecessor, holding the mathematics chair from 1560 to 1567 and from 1582 to 1589, and elected general of his order in 1586, lectured on the first two books of the Tetrabiblios and left in manuscript a commentary on them. The Carmelite Giuliano Ristori

[55] Rutkin, "Astrology", 542.

(1492–1556), who held the chair from 1543 to 1550, also left a manuscript commentary on the *Tetrabiblios*, as well as other works in the occult sciences.[56]

The belief in the influence of the stars was widely held, and astrology of every sort was widely popular, a part of everyday life. As it has been pointed out, such is the significance of astrology that trying to understand the society and culture of early modern Europe without taking astrology into account is as plausible as trying to understand our own time without looking at the influence of economics and psychoanalysis.[57] Even such a shrewd and worldly-wise surveyor of the political scene as the historian Francesco Guicciardini (1483–1540), who publicly ridiculed astrologers and had a reputation for seeing the folly of the whole endeavor, secretly consulted a specialist astrologer who produced a detailed horoscope that Guicciardini took very seriously.[58] This popularity would even continue well into the seventeenth century and beyond. In a letter of July 14, 1642, the eminent disciple of Galileo, Bonaventura Cavalieri (1598–1647), who held the mathematics chair at Bologna from 1629 until his death, lamented to Evangelista Torricelli (1608–1647), another great disciple of Galileo who followed him as mathematician to the Grand Duke in 1642, that there was no public interest—or money, honor, or power—in the mathematical sciences, whereas there was plenty in judicial astrology and that one should keep rigorous science to oneself while adapting to the demands of the majority in making horoscopes and prophecies.[59]

Astrology was also defended by the great Aristotelian philosopher Pietro Pomponazzi in his *De incantationibus* (*On Incantations*), finished in July 1520 and circulated in manuscript though not published until 1556, and *De fato* (*On Fate, Free Will, and Predestination*), written a

---

[56] Charles Schmitt, "The Faculty of Arts at the Time of Galileo", *Physis* 14 (1972): 258–59.

[57] William R. Newman and Anthony Grafton, eds., "Introduction: The Problematic Status of Astrology and Alchemy in Premodern Europe", in *Secrets of Nature: Astronomy and Alchemy in Early Modern Europe* (Cambridge, Mass.: Harvard University Press, 2002), 14.

[58] Anthony Grafton, *Cardano's Cosmos: The Worlds and Works of a Renaissance Astrologer* (Cambridge, Mass.: Harvard University Press, 1999), 10–11.

[59] Eugenio Garin, *Astrology in the Renaissance: The Zodiac of Life*, trans. Carolyn Jackson and June Allen (London: Routledge & Kegan Paul, 1983), 6–7.

few months later in November 1520 and published in 1567. While he was not skilled in astrology, following Aristotle he believed in a great chain of causation beginning and guided by the Unmoved Mover through the lesser realities of the celestial intelligences that guided each heavenly sphere until it affected the sublunary world and determined the course of events here on earth. At a natural level, therefore, and guided by the heavens and the laws of nature, nothing was truly miraculous, nor was individual divine intervention or the intervention of demons and angels possible. The world was governed by an astral determinism ("what happens, happens inevitably"), but this was only at a philosophical level, he argued, and determinism was to be rejected by a Christian belief in divine providence and free will. "Since human wisdom is almost always in error, nor can man attain the genuine truth, especially in relation to divine mysteries, by means of purely natural reasoning, one must in everything stand by the determination of the Church, which is guided by the Holy Spirit."[60]

But toward the end of the fifteenth century Giovanni Pico della Mirandola (1463–1494), the Renaissance humanist and philosopher, set forth the most systematic, expansive, and devastating attack on astrology since antiquity in his posthumous (1496) *Disputationes adversus astrologiam divinatricem* (*Disputations against Divinatory Astrology*).[61] The criticisms of Pico stung and led to many attempts to defend and reform astrology, including those of Girolamo Cardano (1501–1576), a celebrated physician and mathematician, and probably the most famous astrologer of the sixteenth century. He sought to return to the classical sources, to get beyond what he saw as the corrupting Arab influences, and to apply the literary methods of humanism to these sources. He wrote extensively on astrology, drawing up technical treatises as well as horoscopes for the living and the dead, seeing it as a noble and pious response to the beauty, harmony, and regularity of God's creation. As he put it in his dedication to his very substantial *Commentary on Ptolemy's Tetrabiblios*:

[60] Jill Kraye, "Pietro Pomponazzi (1462–1525) Secular Aristotelianism in the Renaissance", in *Philosophers of the Renaissance*, ed. Paul Richard Blum (Washington, D.C.: Catholic University of America Press, 2010), 104–9.

[61] Steven Vanden Broeke, *The Limits of Influence: Pico, Louvain, and the Crisis of Astrology* (Boston: Brill, 2003).

Nothing comes closer to human happiness than knowing and understanding those things which nature has enclosed within her secrets. Nothing is more noble and excellent than understanding and pondering God's supreme works. Of all doctrines, astrology, which embraces both of these—the apotheosis of God's creation in the shape of the machinery of the heavens and the mysterious knowledge of future events—has been unanimously accorded first place by the wise.[62]

Cardano was fully aware of the limits of astrology, considering it more of an art than a science, and the difficulties of making accurate predictions, but he felt sufficiently bold to cast and publish a horoscope of Jesus Christ. While this was not the first time this had been done, and Albertus Magnus thought it permissible insofar as Christ was a man, in the more hostile atmosphere of Pope Pius V, a former inquisitor, he fell afoul of the Inquisition and was arrested, tried, and imprisoned in 1570. Within a few months, and his recantation, he was released but prohibited from teaching and writing. (He was professor of medicine at Bologna at the time.) He retired to Rome, where he wrote the most famous, and certainly the most revealing, autobiography of the Renaissance: *The Book of My Life*, reissued in 2002 by *The New York Review of Books*.

The 1559 Index of Paul IV and the Tridentine Index of 1564 prohibited all books of divination and magic, and all astrological books except those touching upon natural observations that affected navigation, agriculture, and medicine. While they prohibited deterministic predictions, they allowed for predictions where the heavens only inclined not necessitated something. In 1586 the bull *Coeli et terrae*, of Pope Sixtus V, condemned divination in all its forms primarily because knowledge of the future belonged to God alone and all attempts to understand the future were the dangerous products of man's deluded pride. The whole tone of the bull was to emphasize man's lowly deficiency and God's transcendent omnipotence and superiority. Among those condemned were astrologers, called "genethliacs" (casters of nativities), "mathematici", or "planetarians" who headed the list of superstitious and dangerous practitioners.

---

[62] Germana Ernst, "Astrology, Religion, and Politics in Counter-Reformation Rome," in eds. S. Pumfrey, P. L. Rossi, and M. Slawinski, *Science, Culture, and Popular Belief in Renaissance Europe* (Manchester: Manchester University Press, 1991), 252.

While natural astrology for navigation, agriculture, and medicine was permitted, it condemned judicial astrology completely, making no distinction between deterministic and nondeterministic predictions.[63] Perhaps the severity was due to his experience as inquisitor in the Venetian Republic, where the connections between astrology, natural magic, and Aristotelian naturalism (the Paduan Aristotelian Pietro Pomponazzi) were strong.[64]

Even before the Index of 1596 returned to the earlier, less rigid allowance, the bull was being interpreted according to the rules that allowed for nondeterministic predictions, though confusion continued. There were those who were more aggressive against all judicial astrology, especially among the Jesuits, and that was how it stood until Pope Urban VIII's 1631 bull *Inscrutabilis*, which banned political astrology, most particularly in reference to popes and their families.

One area of Galileo's life which has been ignored, and even denied, until recent times despite it being well documented, was his involvement in astrology. That he should practice astrology should not be surprising since that was normal for a mathematics professor into the early seventeenth century. I suppose that it was considered embarrassing by more recent writers that the father of modern science, as some have called him, and the icon of science and modernity, should be involved in a practice that is seen as the antithesis of all this. It is absolutely certain that Galileo was a practicing astrologer during most if not all of his career, that he practiced it extensively, and that he was famous for it, with distinguished people coming to him for horoscopes and predictions, including the family of the Grand Dukes. It is also absolutely certain that he did horoscopes not just for patronage reasons, to curry favor among the great, but also for personal reasons, for his own family (his two daughters) and friends (e.g., Giovanfrancesco Segredo). While some of the most famous charts have not survived, there still remain some twenty-five astrological charts, plus a number of chart analyses. Here is his analysis for his favorite daughter, Virginia.

---

[63] Ibid., 248–51.

[64] Ugo Baldini, "The Roman Inquisition's Condemnation of Astrology: Antecedents, Reasons and Consequences", in *Church Censorship and Culture in Early Modern Italy*, ed. Gigliola Fragnito, trans. Adrian Belton (Cambridge: Cambridge University Press, 2001), 91.

Virginia's character (*De moribus Virginiae*)

First, therefore, Saturn, Mercury and the Moon in separated places, that is, in no aspect between them, indicates a certain discord between the rational [Mercury and Saturn] and sensitive faculty [moon], because Mercury is the strongest and in the sign it rules. But since the Moon [which rules the affects] is weak and found in an obedient sign, reason rules the affects.

Saturn, the significator of mores, since it is exalted, promises that they [the mores] are proper and severe, although mixed with some poison, which, although mitigated and tempered by a beneficial sextile [60 degree] aspect of fortunate Jupiter with a powerful Mercury, makes one, additionally, patient of labors and disturbances, solitary, taciturn, sparing, desirous of one's own advantage, jealous, but not always truthful in promises. Also, a fortunate Sun bestows a certain air of authority and arrogance of manner on a person. Spica rising adds in addition charm and piety. Libra, a human sign, supplies humanity and civility.

Her Mind (*De ingenio*)

With respect to her mind, Mercury endowed with many dignities promises a fine mind. Moreover, since Jupiter is conjoined, it increases wisdom, prudence and humanity. Also, a fortunate and powerful Saturn especially benefits memory. And Libra rising with many planets favors intelligence.[65]

While there is some evidence that he may have turned against astrology late in his life, most of his life he was an avid practitioner.[66] There is no evidence that he publicly taught astrology at Pisa or Padua, but his immediate predecessors certainly did. Medical students needed to know how to do horoscopes so as to ascertain what remedies would be needed, and we know that the majority of his students at Padua were medical students. He may have taught astrology privately at Padua, and his first brush with the Inquisition, in 1604,

[65] H. Darrel Rutkin, "Galileo Astrologer: Astrology and Mathematical Practice in the Late-Sixteenth and Early-Seventeenth Century", *Galilaeana* 2 (2005): 119.

[66] Ibid.; Massimo Bucciantini and Michele Camerota, "Once More About Galileo and Astrology: A Neglected Testimony", *Galilaeana* II (2005): 229–32; and Nick Kollerstrom, "Galileo's Astrology", *The Mountain Astrologer*, accessed January 24, 2017, http://www.skyscript.co.uk/galast.html#7.

included the accusation that he practiced a deterministic astrology that denied human free will. If he did not teach astrology, it perhaps might be because Clavius himself had rejected the teaching of astrology as part of mathematics in Jesuit institutions; and that he did not talk about astrological theory and write about it publicly might be due to the condemnations of Pope Sixtus V in 1586 and Clavius' rejection of it.

Certainly, interest and belief in astrology remained very strong among many churchmen, including Pope Urban VIII. In 1630, while Galileo was trying to get approval for his *Dialogue on the Two Chief World Systems* and a few years before his own trial, a major eruption occurred in Rome when the abbot of the monastery of Santa Prassede and onetime general of the Vallombrosian order, Orazio Morandi—an old friend of Galileo and the most sought-after astrologer in Rome—was arrested with his monks and put on trial for predicting the imminent death of Pope Urban VIII. We will discuss this in more detail in the next chapter.

Chapter Four

Galileo Triumphans: Galileo as Academic Luminary

Galileo's triumphant progress in Rome and his well-nigh universal acclaim should not blind us to the fact that there were still many who opposed him, and not just because of envy or inertia, though we should never discount those, especially now that he was speaking about more than just the mountains and valleys of the moon or the satellites of Jupiter. To say that the earth was moving and not the center of the universe, and that the sun sat motionless at the universe's center, as he now was starting to say (especially after he saw the phases of Venus), was contrary not just to millennia of learned men but to common sense itself, which had no experience of such a reality. And while there were those who opposed him because of his undermining of the Aristotelian philosophical worldview, it was theology that ultimately was to give him the most trouble. It was his response to theological attacks that would drag him into the dangerous thickets of Catholic biblical scholarship and would trigger the intervention of the Roman Inquisition.

The Council of Trent defined the nature of Sacred Scripture and of its relationship to Sacred Tradition and the Church. In the Fourth Session, on April 8, 1546, the council defined the list of the approved books of Scripture, the authenticity of the Latin Vulgate edition of the Bible despite critical awareness of its limitations, and two other points of interpretation. The first of these was the need for Sacred Tradition. The council stated that the Gospel, first promulgated by Christ and preached by the apostles, was contained both in the written books of Scripture and "in the unwritten traditions which, received by the Apostles from the mouth of Christ himself or from the Apostles themselves, the Holy Spirit dictating, have come down to us, transmitted at it were from hand to hand". It also stated that since God was as much the author of these unwritten traditions as of Scripture, these traditions, "preserved in the Catholic Church in unbroken succession", should be received with a "feeling of equal piety and reverence" as Scripture. The second point of interpretation was the need of the Church (and for Church one really should say its leaders, the bishops, and theologians) for a proper understanding of Scripture: "Furthermore, to control petulant spirits, the Council decrees that, in matters of faith and morals pertaining to the edification of Christian doctrine, no one, relying on his own judgment and distorting the Sacred Scriptures according to his own conceptions, shall dare to interpret them contrary to that sense which Holy Mother Church, to whom it belongs to judge their true sense and meaning, has held and does hold, or even contrary to the unanimous agreement of the Fathers, even though such interpretations should never at any time be published."[1] In these decrees we see not only a clear rejection of the Protestant doctrine of *sola scriptura*, that Scripture alone is the source of revelation and salvation, and so the denial of Tradition, but also of the Protestant idea of private judgment with its rejection of the institutional Church as a necessity in the process of understanding the Gospel.

The theological opposition to Galileo began early, both in Rome and in Florence. As we already know, on May 17, 1611, at a regular meeting of the Inquisition, with Cardinal Bellarmine attending, it was asked if Galileo had been mentioned in any of the proceedings against Cremonini. Later that year, on December 16, Galileo's friend, the painter Lodovico Cigoli, wrote him that a group of people had been meeting regularly at the house of Alessandro Marzimedici (1557–1630), the archbishop of Florence, and were plotting against him because of his views on the earth's motion. The first public attack, however, only came on All Souls' Day 1612, when a Dominican friar, Niccolò Lorini (b. 1544), spoke out against Galileo and the Copernican doctrines as a violation of Scripture. In a letter to Galileo, Lorini denied that he had attacked Galileo, writing that

[1] Richard J. Blackwell, *Galileo, Bellarmine, and the Bible* (Notre Dame, Ind.: Notre Dame University Press, 1991), 181–84.

he had only said a couple of words to the effect "that this opinion of Ipernicus [sic]—or whatever his name is—was contrary to Holy Scripture."[2] The group opposed to Galileo was dubbed the *Lega del Pippione* ("Pigeon League") after their leader Lodovico delle Colombe (b. 1565), whose name means "of the doves". Lodovico—philosopher, poet, and member of the Florentine Academy—had opposed Galileo since the nova of 1604, and in his *Against the Earth's Motion* (1611) he attacked the Copernican view, primarily with philosophical arguments but also theological ones. The Pigeon League was centered on Lodovico; his brother Raffaello (1557–1627), a Dominican who would preach against the Copernicans even from the Duomo, the great cathedral of Florence; and a number of other Florentine Dominicans, of which the most significant were Lorini and Tommaso Caccini (1574–1648).

Among the things that Galileo showed off in Rome during his triumphal visit were sunspots. While sunspots were nothing new, and recorded observations went back at least to 165 B.C. and the Chou dynasty in China, in the West they were generally ignored. First, because they were hard to see with the naked eye due to the brightness of the sun; second, because they were not looked for since there was a presupposition that the heavens were perfect, unchanging, and incorruptible. In A.D. 807, however, one was so large and lasted for so long (eight days) that it could not be ignored, but it was described as being Mercury transiting (crossing) the sun. It is not entirely clear who first observed sunspots with a telescope. The English astronomer Thomas Harriot and the Frisians Johannes and David Fabricius saw them around the end of 1610, with the first to publish being Johannes Fabricius, whose *De Maculis in Sole Observatis* (*On the Spots Observed in the Sun*) appearing in the autumn of 1611, though this work seems to have gone unnoticed. Christoph Scheiner (1573–1650), a Jesuit professor of mathematics and Hebrew at the University of Ingolstadt in Bavaria, first observed them in March 1611 and published on them in January 1612 in three letters to the German Catholic banker Mark Welser under the pseudonym "Apelles". Scheiner saw them as satellites orbiting the sun since he believed the sun to be perfect and

[2] Jerome J. Langford, *Galileo, Science and the Church* (Ann Arbor: University of Michigan Press, 1992), 51.

immutable.[3] Galileo seems to have seen them earlier, for his Venetian friend Fra Fulgenzio Micanzio later clearly recalled him showing them to him in the summer of 1610.[4]

In 1612 Galileo wrote his own three letters to Welser, where he argued that they were on or near the surface of the sun, and therefore could not be satellites but more like clouds. He also for the first time unequivocally and publicly endorsed Copernicanism. Also in 1612 (in July) Galileo wrote to Carlo Cardinal Conti in Rome, asking about issues raised by sunspots, particularly wondering if the Aristotelian worldview was compatible with Scripture. Conti replied that he thought the Scriptures did not favor the Aristotelian view of the incorruptibility of the heavens and that it was the common opinion of the Fathers of the Church that they were corruptible. That the earth moves through the heavens, as the Pythagoreans and Copernicus believed, appeared less agreeable he thought since we would then have to interpret those places that speak about the earth staying still as just a common and uneducated manner of speaking, something that should not be admitted without great necessity. He did, however, point out that Diego Zuñiga in his *Commentary on Job* had stated that it was more agreeable to Scripture that the earth move, though this was not the common opinion. It is interesting that when the *Letters on Sunspots* was published in 1613, the censors demanded that all references to Scripture as supporting Galileo's position be taken out, but that his support of Copernicanism could stay.

The year 1613 would also see the occurrence that would set in motion a chain of events that would lead to the 1616 condemnation of Copernicanism. It began on December 12, 1613, when Dom Benedetto Castelli (1578–1643)—a Benedictine monk who had studied under Galileo at Padua, becoming his good friend and protégé, and now professor of mathematics at Pisa (since 1613)—breakfasted with the Grand Duke, his wife, and his mother, the Dowager Grand Duchess Christina of Lorraine. Asked about Galileo's telescope, his

---

[3] Albert Van Helden, "Galileo and Scheiner on Sunspots: A Case Study in the Visual Language of Astronomy", *Proceedings of the American Philosophical Society* 140 (1996): 358–96; and William R. Shea, "Galileo, Scheiner, and the Interpretation of Sunspots", *Isis* 61 (1970): 488–519.

[4] David Wootton, *Galileo: Watcher of the Skies* (New Haven, Conn.: Yale University Press, 2010), 125–26.

discoveries, and his opinion of the motion of the planets, Castelli defended the validity of Galileo's discoveries, and in his later letter to Galileo he wrote that he thought that he had handled things well and had received a favorable response from the group. Among those attending were Don Antonio de' Medici (1576–1621), son of Francesco I by his then mistress Bianca Capello, and Cosimo Boscaglia, a professor of philosophy at the University of Pisa. Boscaglia spent much of the time whispering into the Dowager Grand Duchess' ear, and while Boscaglia admitted that Galileo's discoveries were real, he thought that the earth's motion was incredible and impossible, especially as it was opposed to Scripture. After the meal Castelli was called back to the Dowager Grand Duchess' chambers, where he found not only her but the Grand Duke, his wife, Don Antonio, Professor Boscaglia, and Don Paolo Giordano Orsini (1591–1656), a member of the ancient and powerful Roman Orsini, a relative of the Medici Grand Duke through his mother, and the brother of Alessandro Orsini to whom Galileo would dedicate his *Discourse on Tides* in 1616. The Grand Duchess then began to argue against Galileo's views from Sacred Scripture, mentioning in particular the passage from the Book of Joshua (10:12–14) about God stopping the motion of the sun. Castelli defended Galileo's views and believed that the Grand Duke, the Archduchess, Don Antonio, and Don Paolo Giordano were on his side. Even the Grand Duchess, he thought, did not seem to be really opposed to him, but had argued so as to hear him out. Boscaglia, however, remained silent throughout the two-hour discussion.

After Castelli wrote to him about this, Galileo responded with his famous *Letter to Castelli* (December 21, 1613), where he enunciated his principles of biblical interpretation in reference to science.[5] These seem to have been based solely on his intelligent common sense since he quotes no authorities in his defense. First, he conceded that Scripture could not err though some of its interpreters could, very frequently by reading it too literally. This could not only lead to contradictions but even to serious heresies and blasphemies since it would be necessary to attribute to God "both feet and hands and eyes, and not least bodily and human feelings such as anger, repentance,

[5] Thomas F. Mayer, ed., *The Trial of Galileo 1612–1633* (North York, Ontario: University of Toronto Press, 2012), 49–55.

hatred, and sometimes also forgetfulness of the past and ignorance of the future".[6] In Scripture, many things were accommodated to the incapacity and limitations of the common people, and so needed an interpretation that was different from the apparent literal meaning. That was why, he wrote, in disputes over natural phenomena Scripture should be reserved to the last place. Both Scripture and nature were derived equally from the divine Word, the former by the dictation of the Holy Spirit, and the latter by divine command. Scripture was adapted to the understanding of all people, to the capacity of unrefined and undisciplined peoples, but nature, being inexorable and immutable, was revealed to us by our sensory experience or necessary demonstrations. Once something was proved by clear sensory experience or necessary demonstrations, particularly in things very far from the primary function of Scripture (those articles of faith which are necessary for salvation) or in things spoken of only incidentally or fragmentarily of the earth, sun, or other creatures (physical conclusions), then Scripture should give way since two truths could not contradict each other. Even more strongly, he thought that it would not be prudent to oblige anyone to hold a specific meaning of a scriptural passage where the contrary could ever be proved by the senses and demonstrative and necessary reasons. For Galileo, the human mind was the greatest gift of the Creator and one that should be respected.

> And who would want to put a limit to human understanding? Who would want to assert that everything that is knowable in the world is already known? And for this reason, besides the articles concerning salvation and the establishment of the faith, against the firmness of which there is no danger whatever that could arise against a valid and effective teaching, *the best advice would perhaps be to avoid adding others unnecessarily....* But I do not think it is necessary to believe that the same God who has given us senses, discourse and intellect *wanted to leave aside their use* and to give us with other means the notice of things that through them we can grasp, and especially in those sciences of which one reads in scripture a tiny particle and variant conclusions, such as precisely is astronomy, of which there is such a small part that we cannot even find the names of the planets mentioned.[7]

[6] Ibid., 50.
[7] Ibid., 52; emphasis in original.

Galileo particularly criticized those who would bring scriptural passages ("that irresistible and terrible weapon, the mere sight of which terrifies even the most skillful and expert champion")[8] into disputes about natural phenomena. He also criticized those who met experiments and necessary demonstrations with sophisms, paralogisms (pieces of illogical reasoning), and fallacies. He concluded his letter with a discussion of how if one insisted on a literal reading of that passage in Joshua, it agreed very well with the Copernican system, since, strictly speaking, in the system of Ptolemy and Aristotle, if you stop the sun, you shorten not extend the day.

On December 21, 1614, in the Church of Santa Maria Novella, one of the main churches in Florence and famous for its intricate classical façade by Alberti, Tomasso Caccini, reader of Sacred Scripture for that year there and preaching on the Book of Joshua, came to that classic passage against heliocentrism and used it as an opportunity to attack Copernicanism, the supporters of Galileo (the Galileists), and mathematicians (another name for astrologers) in general, seeing their views as contrary to Scripture. Supposedly, and this story can only be traced to the late eighteenth century, he began his sermon by quoting from Acts 1:11: "Viri Galilaei, quid statis adspicientes in coelum?" ("Ye men of Galilee, why do you stand looking up into heaven?"), an allusion to Galileo and his followers. Luigi Maraffi, a Florentine Dominican and onetime prior of Santa Maria Novella, and a great friend of Galileo, would later write apologizing for Caccini's behavior. Unfortunately for Galileo, Maraffi, who became an assistant to the Secretary of the Index, Francesco Maddaleni Capiferro, and would later become a consultor to the Index in May 1616, would die later in 1616.

On January 12, 1615, Prince Cesi wrote Galileo that he should be cautious and leave the defense to others since Cardinal Bellarmine had told him that he thought Copernicus' opinion was heretical and without a doubt opposed to Scripture. Cesi also remarked that it was very easy to have a book prohibited or corrected, and that it was done even in doubtful cases. "Telesio and Patrizi were prohibited, and when the others are not ready, this reason never lacks, that they are advanced books and too much to be read safely; and the contrary

[8] Ibid., 53.

are most hateful to Aristotle."[9] He also feared that if Copernicus were brought to the Congregation of the Index, he would be prohibited.

On February 7, 1615, another Florentine Dominican, Niccolò Lorini (b. 1544), sent a letter attacking Galileo to the Dominican Paolo Cardinal Camillo Sfrondati (1560–1618), prefect of the Congregation of the Index, who passed it on to the Inquisition. In it he wrote how, "moved by nothing but zeal" and occasioned by Caccini's sermon on chapter 10 of Joshua, he felt that he had to tell them of those who were expounding Holy Scripture in their own way and against the common exposition of the Fathers of the Church and against Holy Scripture itself. While it was common duty of all Christians to oppose heresy, he thought it was particularly a duty of all theologians and preachers (of which he was the lowest of all) and of all Dominican friars ("the black and white hounds of the Holy Office") to do so. He claimed that a letter, which he had enclosed, had been passed through everybody's hands here in Florence. He wrote that it originated from those known as "Galileists" who, following Copernicus, "affirm that the earth moves and the sun stands still". His Dominican confreres thought it full of suspect and rash propositions, which he had underlined, such as "that certain ways of speaking in the Holy Scripture are inappropriate; that in disputes about natural effects the same Scripture holds the last place; that its expositors are often wrong in their interpretations; that the same Scripture must not meddle with anything else but articles of faith; and that, in questions about natural phenomena, philosophical or astronomical argument has more force than the sacred and divine one." He also wrote that he had heard that they spoke disrespectfully of the Fathers of the Church and Saint Thomas, trampled underfoot Aristotle's philosophy, and, trying to appear clever, spread a thousand impertinences around the city.[10]

On February 16, Galileo, fearful that his opponents would use an inferior copy of his *Letter to Castelli* against him, wrote Monsignor Piero Dini (1570–1625), a Florentine working in the Roman Curia, a nephew of Cardinal Bandini (with whom he lived), and a friend of Galileo, concerning Caccini's attack and the machinations of others. He also asked him to see Father Grienberger, "a very good

---

[9] Ibid., 59.

[10] Maurice A. Finocchiaro, *The Galileo Affair* (Berkeley: University of California Press, 1989), 134–35.

friend and patron of mine",[11] and, if he thought it appropriate, Cardinal Bellarmine, "whom these Dominicans seem to want to rally around" with the hope of condemning Copernicanism, should be shown a correct copy of the letter which he had enclosed. Playing upon the recent conflict between the Dominicans and Jesuits in the *De auxiliis* controversy, he told Dini that his most immediate act should be to approach the Jesuit Fathers whose knowledge was much above the common education of the Dominican friars. He also mentioned that he had written a much longer work on the topic, what would become his *Letter to the Grand Duchess Christina*, which he could send when he had finished polishing it.

On February 25, 1615, the Inquisition met at the residence of Cardinal Bellarmine, with six other cardinals and other officials attending, and they examined Lorini's letter and his copy of Galileo's *Letter to Castelli*. It was decided to secure a copy of the original from the archbishop of Florence. Cardinal Millini (1562–1629), Secretary of the Inquisition since 1612 and highly regarded by Pope Paul V, who had made him a cardinal, wrote to the archbishop of Pisa, where Castelli taught, requesting to see the original letter. In a letter to Galileo of March 12, 1615, Castelli wrote of his being called to a meeting with the archbishop of Pisa: "He took me to his office, seated me, and began to ask after your health. I had scarcely finished answering when he began to exhort me to give up certain extravagant opinions, and particularly that of the Earth's motion. He said that this was for my own good and that he meant me no harm because these opinions, in addition to being silly, were dangerous, scandalous, and rash, being directly contrary to Scripture." He told Castelli that these opinions had been Galileo's ruin and, getting angrier, that it would soon be known by all that his ideas were silly and deserved condemnation. The archbishop then asked for a copy of the letter from Galileo, and when Castelli replied that had no copy of it, he was asked to write to Galileo to get one. Castelli hoped that that would be enough.[12]

Dini replied on March 7, 1615, telling him that he had shown the letter to many including Grienberger and Cardinal Bellarmine, with

[11] Ibid., 55.

[12] William Shea and Mariano Artigas, *Galileo in Rome: The Rise and Fall of a Troublesome Genius* (Oxford: Oxford University Press, 2003), 63–64.

whom he spoke at length. Bellarmine told him that he did not think that Copernicus would be condemned, the worst being some sort of appended note to his doctrine just to save appearances. But if an attempt were made to claim it as a real description of the universe, there would be significant problems in reference to Scripture which attributed motion to the sun. Bellarmine also told him that he would gladly read Galileo's work when it was ready and that he would consult with Grienberger. Grienberger, clearly worried about opposing scriptural passages, told him that he would have preferred that Galileo carry out his demonstrations before getting involved with Scripture, and that his arguments in favor of his position were more plausible than true. In a letter of March 14, Dini wrote to Galileo that Cardinal Barberini had told him that Galileo should speak cautiously and only as a professor of mathematics—that is, stay away from Scripture and theology—and that Barberini had not heard any rumors against Galileo.

In an earlier letter (February 28, 1615), Giovanni Battista Ciampoli, a friend of Galileo living in Rome and a confidant of Cardinal Barberini, wrote that Galileo's fears were exaggerated; that no one he had spoken to, including Galileo's friends Monsignor Dini and Fra Luigi Maraffi, had seen any danger to Galileo from the attacks of the Dominicans in Florence. He also passed on to him Cardinal Barberini's advice that since theologians claim Scripture for their own, it would be wise for him to stick to physical and mathematical limits and not to depart from the arguments of Ptolemy and Copernicus.

Ciampoli (1589–1643), who would play such a great part in the Galileo Affair, was born in Florence from an old and noble but impecunious family, studied with the Jesuits and Dominicans, and by his extraordinary linguistic and literary ability gained an entrée into important Florentine literary circles and that of Grand Duke Ferdinando I, becoming a friend of the sons of Prince Cosimo. Promoted and supported by the Florentine aristocrat and poet Giambattista Strozzi (1551–1634), who was also a patron of Galileo, he was able to further his career and studies. His meeting with Galileo in 1608 at the Grand Duke's villa changed the trajectory of his life, and he became enamored of mathematics and a strong supporter of Galileo's work. He studied at Padua, Bologna, and finally at Pisa, where he received a degree in canon and civil law in 1614. The same year he became a cleric, so as to be able to receive Church benefices and pursue his career in the Roman Curia. In 1621 he would be appointed Secretary of Briefs by Pope Gregory XV, and in 1623 he would become Secret Chamberlin to Pope Urban VIII, whom he had known as Cardinal Barberini and with whom he would become relatively close. Ciampoli would become one of Galileo's most important connections in Rome, and in 1618 he would join the Academy of the Lynxes.

On March 7, 1615, Prince Cesi sent Galileo a recently published book by the Carmelite friar Paolo Antonio Foscarini (c. 1565–1616), which Cesi thought could not have come at a better time for their cause. Published in Italian, this pamphlet-sized *Letter concerning the Opinions of the Pythagoreans and of Copernicus* was the first published work defending Galileo and showed how Copernicanism was not contrary to Scripture. Foscarini, born in Calabria and twice provincial there, was well-versed in mathematics, natural philosophy, and theology (where he taught it at the University of Messina), and was in the process of writing a seven-volume encyclopedia of knowledge. That year he was also preaching a Lenten series in Rome. While Galileo would remain primarily on the safer level of general principles, Foscarini grasped the nettle and dealt not only with general principles but directly with key passages of Scripture. While he realized the boldness of his work and that Copernicanism had yet not been proven true, he believed that it might be shown to be true in the future and, by showing how Scripture could be consistent with it, scandal and embarrassment could be avoided. He later sent to Bellarmine, perhaps in April, the *Letter* and a short *Defense* of his letter in Latin where he listed the many places where his general principles could be found in the Fathers, in the scholastics such as Saint Thomas Aquinas, and in more recent theologians.

Galileo had promised Dini to send a more extensive treatise than his *Letter to Castelli*. In Antonio Favaro's National Edition of Galileo's works the *Letter to Castelli* takes up only seven pages, while his more complete treatise, the justly famous *Letter to the Grand Duchess Christina*, takes up thirty-nine pages. The letter was finished in midsummer, and it was somewhat widely copied and circulated (Favaro examined thirty-four manuscript copies for his edition), but it does not seem to have been seen by those Roman authorities who condemned Copernicanism some months later and was only formally published in 1636 in Protestant Strasburg in a small book with the Italian text in the right column of each page and a Latin translation done by Galileo's friend Elia Diodati, under a pseudonym, in the left column.[13]

Galileo was no theologian, and his library showed no extensive reading of theology, especially the works of the Fathers of the Church and the theologians whom he quotes, but with the aid of others, particularly Castelli, he gathered many citations, especially from Saint Augustine, to give authoritative support to the general points we have already seen in his *Letter to Castelli*. In some respects this was a weaker text than before insofar as he wrote, also following Augustine, that where science was not *certain* about a fact of nature then the literal interpretation should prevail "inasmuch as divine wisdom surpasses all human judgment and speculation". However, he also undercut that by writing that since Scripture was so often ambiguous it would be "very prudent not to allow anyone to commit and in a way oblige scriptural passages to have to maintain the truth of any physical conclusions whose contrary could ever be proved to us by the senses and demonstrative and necessary reasons".[14] He thought that this principle should be applied even where Scripture was unanimous, and the Fathers were unanimous in their understanding of it since he doubted that the consensus of Fathers alone could make any conclusion about natural phenomena an article of faith; nor did this seem to be in accord with what the Council of Trent had decreed about the consensus being limited to matters of faith and morals.[15] It was the opinion of the Fathers, he believed, that the sacred authors of Scripture had deliberately refrained from discussing these topics because they were not ordered to the salvation of souls. He also quoted the great Church historian Cardinal Baronius (1538–1607) that "the intention of the Holy Spirit is to teach us how one goes to heaven and not how heaven goes."[16]

---

[13] Ernan McMullin, ed., *The Church and Galileo* (Notre Dame, Ind.: Notre Dame University Press, 2005), 265–78; Annibale Fantoli, *Galileo for Copernicus and for the Church*, trans. G. V. Coyne, 2nd ed. (Vatican City: Vatican Observatory, 1996), 189–208; Richard J. Blackwell, *Galileo, Bellarmine, and the Bible* (Notre Dame, Ind.: Notre Dame University Press, 1991), 75–85; and Jean Dietz Moss, "Galileo's Letter to the Grand Duchess Christina: Some Rhetorical Considerations", *Renaissance Quarterly* 36, no. 4 (1983): 547–76.

[14] Finocchiaro, *Galileo Affair*, 96.

[15] Ibid., 104.

[16] Ibid., 92, 93, 94, 96.

While Galileo speaks of the importance of demonstrations some twenty-five times in the text, he himself offers no demonstration of the scientific truth of Copernicanism. He refers to the fact that he has confuted Ptolemy's and Aristotle's arguments, to the "truths" that he has pointed out which have convinced those most competent in astronomical and physical science. He refers a number of times to those who in "their simulated religious zeal" use Scripture against him, "to shield the fallacies of their arguments".[17]

He also quoted from Copernicus' own preface where he stated that mathematics was written for mathematicians and defended himself against those who might attack him, quoting Scripture against him. Finally, Galileo ended the work showing how the Joshua passage so often used against the heliocentric theory worked better with that than with the Ptolemaic system.

To call your opponents stupid and hypocritical—even if it is true in some cases—is not perhaps the best way to persuade people, however satisfying it may be to one's wounded sense of honor. In Galileo's defense, however, being called a heretic could not only be hurtful to one's pride but could also lead to one's imprisonment or death and the ruin of one's work. While Galileo could complain of theologians trespassing on his territory as a scientist, the theologians could have complained just as much about Galileo trespassing on theirs. He, who was neither a bishop nor a theologian, had made himself an interpreter of Scripture. That his understanding of Scripture was in fact correct, supported by the Tradition and later theologically vindicated by Pope Leo XIII's encyclical on Sacred Scripture *Providentissimus Deus* (1893), is not unimportant, but that is after the fact. At this time, it was a dangerous game to play with Scripture, and Galileo had been warned of this, but Galileo was not one to underestimate his powers of persuasion. Rallied by Foscarini's pamphlet, the situation seemed to him too precarious and the stakes far too great not to do something, and passivity was never part of his personality.

On March 20, 1615, Caccini—now residing at the Dominican convent of Santa Maria sopra Minerva in Rome, having gained his bachelor of sacred theology—came before another Dominican, Michelangelo Seghizzi (1565–1625), the Commissary General of the

---

[17] Ibid., 89, 90.

Holy Office, the leading official of the Inquisition. Seghizzi, who would only serve as Commissary General for eighteen months before being made bishop of Lodi, had joined the Dominicans at fourteen, served as an inquisitor in Lombardy from 1603 until 1615, where he generated a great deal of business for the Inquisition in Rome, and according to his epitaph was "a hammer of heretics".[18] In his deposition, Caccini charged Galileo with holding that the sun was motionless at the center of the world, with the earth moving as a whole as well as with diurnal motion—propositions he thought repugnant to the divine Scripture expounded by the Holy Fathers and to the faith. It was not just others, he said, who told him about these propositions of Galileo, but he had read them in Galileo's own book on sunspots. He also recounted that he had been told by another Dominican, Fra Ferdinando Ximenes, the regent at Santa Maria Novella, that some of the disciples of Galileo believed that God was not a substance but an accident (Aristotelian terms, the first signifying the essence of something and the second signifying a secondary reality that inheres in a substance), that God was sensuous since he has divine senses, and that miracles made by the saints were not real miracles. He also mentioned that while many regarded Galileo as a good Catholic, others regarded him with suspicion since he was in contact with Germans (understood to be the home of heresy) and was very close to Fra Paolo Sarpi, "so famous in Venice for his impieties", with whom he was still exchanging letters.[19]

At the Thursday, April 2, meeting of the Inquisition, the pope, attended by seven cardinals, asked that Caccini's deposition be forwarded to the Florentine inquisitor and asked that Attavanti and Ximenes be interrogated. In the deposition taken from Ximenes by the Florentine inquisitors (November 13, 1615), Ximenes confirmed what Caccini had said about the opinions of Galileo's disciples, except for the part about saints' miracles not being real miracles, and in particular named a certain Florentine nobleman and cleric, Giannozzo Attavanti. He stated that while Attavanti had been discussing problems of conscience with Ximenes in Ximenes' room at Santa Maria Novella, the conversation had turned to Caccini's

---

[18] Thomas F. Mayer, *The Roman Inquisition: A Papal Bureaucracy and Its Laws in the Age of Galileo* (Philadelphia: University of Pennsylvania Press, 2013), 112–18.

[19] Finocchiaro, *Galileo Affair*, 136–43.

sermons, particularly the one discussing how the sun stood still. Ximenes thought that Attavanti was expressing more Galileo's opinions than his own, and while he could not be sure whether Attavanti was speaking argumentatively rather than categorically, he rebuked him for these false and heretical ideas.

When the Florentine inquisitor took Attavanti's deposition the next day, he confirmed that he studied casuistry with Ximenes, and while not a student of Galileo's, he had spoken with him but had never heard him say anything that conflicted with Scripture or the Catholic faith. He, however, had heard him support heliocentrism and referred to Galileo's *Letters on Sunspots*. He had also heard him argue about the Joshua passage where he admitted that the sun stopped miraculously "but nevertheless (viewed from the outside) it does not move with a progressive motion". In reference to the points about Galileo saying that God was an accident and so on, he replied that when he was in Ximenes' room they were discussing topics in Saint Thomas Aquinas "in the manner of a disputation, and not otherwise", and that Caccini, overhearing the discussion, must have imagined that these were assertions of Galileo—which they were not. He believed this was what likely happened since in an earlier incident, in July or August 1613, Caccini, in his room, had overheard another discussion, this time about Copernicus' opinions, and had left his room and interrupted them to assert that these opinions were heretical and that he wanted to preach about it from the pulpit. When asked about Galileo's faith, Attavanti said that he believed him to be "a very good Catholic; otherwise he would not be so close to these Most Serene Princes".[20]

We have already mentioned many times the figure of Robert Bellarmine (1542–1621), and he looms large above all of these events; and as a leading cardinal inquisitor and the most respected theologian in the Catholic world, and well known outside it, involved in every major theological dispute of the day, often very publicly, and respected as a saintly man, his actions would be of decisive importance in all that followed and without which there would be no Galileo Affair.

[20] Thomas Companella, *A Defense of Galileo, the Mathematician from Florence*, trans. Richard J. Blackwell (Notre Dame, Ind.: University of Notre Dame Press, 1994), 143–46.

Bellarmine was the third son of twelve children from an old and distinguished but impoverished Tuscan family at Montepulciano, an ancient town forty-three miles southeast of Siena. He was also a nephew on his mother's side of the great religious reformer Marcello Cervini, who briefly became Pope Marcellus II (1555). Highly intelligent and lively, raised in the shadow of the Reformation and Counter-Reformation, having entered the Jesuits in 1560 despite his father's wish for him to follow the more remunerative path of medicine, Bellarmine early on absorbed the intensely zealous character of the Jesuits in their first heroic decades. While his whole life was spent in the promotion and defense of the Catholic faith, it was never at the cost of truth or charity, and his work was remarkably free of the gratuitous name-calling and mudslinging that disfigured so much of the polemical literature of the time. After teaching at Louvain in the theologically divided Low Countries (1569–1576), he was brought to the Roman College to become the professor of controversial theology, delivering lectures to the elite of the Jesuit and Catholic world on the errors of heresy, as Protestantism certainly appeared to him. It was from these lectures that we get one of the monuments of Catholic theology, his three-volume work *Disputationes de controveriis christianae fidei, adversus huius temporis haereticos* (*Disputations about the Controversies of the Christian Faith Against the Heretics of this Time*)—or, as it is generally known, the *Controversies*. It breathes the militant spirit of the Counter-Reformation, and one can have no doubt about his view of heresy by just looking at its preface: "Let me say this one thing; the perversity of heretics is as much worse than all other evils and afflictions as the dreadful and fearful plague is worse than the more common diseases."[21]

A master of historical and scholastic theology, and a theology more historical than philosophical, Bellarmine became the major theological figure in Rome, as erudite as he was holy, becoming rector of the Roman College (1592–1595) and being made a cardinal in 1599. By the time Galileo met him in 1610, he was a member of five different Roman congregations, including the Inquisition and the Index.

---

[21] Richard S. Westfall, *Essays on the Trial of Galileo* (Notre Dame, Ind.: University of Notre Dame Press, 1989), 6.

As we have seen, he had been involved in the notorious case of the errant philosopher Giordano Bruno, both as a consultor and as a cardinal inquisitor who condemned him to be burned alive at the stake in the Campo de' Fiori. While there was no greater defender of the papacy, his theory that the pope's power over secular princes was only indirect was held against him by the more extreme papalists (who thought him heretical because of it), and Pope Sixtus V put the first volume of his *Controversies* on the Index, though the pope's death before the publication of the new Index allowed for its removal before there was any public embarrassment.

While he became the main theological advisor to Pope Clement VIII and was made a cardinal by him, he fell out of favor and was sent away in 1602 to become a model archbishop at Capua in the Kingdom of Naples. He only returned in 1605 for the two papal elections of that year (Leo XI only reigned twenty-six days and Bellarmine was a leading candidate in both elections) and was asked by the new pope, Paul V, to remain in Rome and allowed to resign his archbishopric. He defended the papacy against Venice during the Interdict crisis of 1606 (where Sarpi used Bellarmine's opinions on the papacy against him) and against King James I of England in 1607–1609. During Galileo's triumphant visit to Rome in 1610, Galileo met with Bellarmine and showed him his discoveries through his telescope. Not long after that Bellarmine wrote to the Jesuit astronomers at the Roman College asking if Galileo's points were correct and was confirmed in their basic truthfulness.

He accepted, as did Galileo, the decisions of the Council of Trent in reference to Scripture and its interpretation. His view of Scripture, however, was far more literal. This stress on the literal meaning of Scripture is likely a product of his controversies with Protestants where anything but a more literalist understanding is useless for argumentation. For Bellarmine, Scripture could not only not err in areas of faith and morals (as traditionally understood), in those things that are necessary to know for salvation, but it could not err even in the smallest point or fact.[22] As he wrote to Foscarini (April 12, 1615): "Nor can one reply that this is not a matter of

[22] Blackwell, *Galileo, Bellarmine, and the Bible*, 32.

faith, because even if it is not a matter of faith 'because of the subject matter,' it is still a matter of faith because of the speaker. Thus
anyone who would say that Abraham did not have two sons and
Jacob twelve, would be just as much of a heretic as somone who
would say that Christ was not born of a virgin, for the Holy Spirit
has said both of these things through the mouth of the Prophets
and Apostles."[23] This view of Scripture led him to create his own
strongly non-Aristotelian biblical cosmology as we have seen. While
he admitted that Copernicanism saved all the appearances better, to
affirm its true reality would be "a very dangerous thing, not only because it irritates all philosophers and scholastic theologians", but
because it would harm the Catholic faith by showing Scripture
to be false. He did, however, accept Augustine's dictum that where
there was a clear proof of some fact of nature, Scripture must give
into it. Thus, in this same letter to Foscarini, Bellarmine wrote that
"whenever a true demonstration would be produced that the sun
stands in the center of the world and the earth in the third heaven,
and that the sun does not rotate around the earth but the earth
around the sun, then at that time it would be necessary to proceed
with great caution in interpreting the Scriptures which seem to be
contrary, and it would be better to say that we do not understand
them than to say that what has been demonstrated is false. But I do
not believe that there is such a demonstration, for it has not been
shown to me".[24] While Bellarmine was a man of intelligence and
charity, he was also a man who had spent his whole life in combat
for an ideal, a man for whom novelty was another name for heresy,
and a man well past his prime, being seventy-three when Galileo
made his third visit to Rome. As Giorgio de Santillana once wrote,
it was unfortunate that Galileo met in Rome not a young Aquinas
but an old Bellarmine;[25] or, more sympathetically, as Westfall put it
about Bellarmine, he was a captive, as each of us is a captive, of his
lifetime's experience.[26]

---

[23] Ibid., 266.

[24] Ibid.

[25] Giorgio de Santillana, *The Crime of Galileo* (Chicago: University of Chicago Press, 1955), xi.

[26] Westfall, *Essays on the Trial of Galileo*, 23.

Bellarmine's letter to Foscarini began by commending him and Galileo for speaking suppositionally and not absolutely as he believed Copernicus spoke. While Bellarmine may not have been aware that the instrumentalist preface (that the heliocentric theory was just a useful fiction) of *De revolutionibus* was not by Copernicus, he must have been aware that both Foscarini and Galileo thought in more realist terms; and, it is quite possible, as some believe, that he was not only giving them a subtle warning but also opening a path for them to continue their work. Galileo, however, did not see it that way. In his reflections on Bellarmine's letter to Foscarini (which he obtained), Galileo wrote in regard to these so-called philosophers that "if they were true philosophers, namely lovers of truth, they should not get irritated, but, learning that they were wrong, they should thank whoever shows them the truth; and if their opinion were to stand up, they would have reason to take pride in it, rather than being irritated. Theologians should not get irritated because, if this opinion were found false then they could freely prohibit it, and if it were discovered true then they should rejoice that others have found the way to understand the true meaning of Scripture and have restrained them from perpetrating a serious scandal by condemning a true proposition."[27]

While the Inquisition's consultant cleared Caccini's version of Galileo's *Letter to Castelli* of any heresy despite a certain improper way of speaking, at a meeting of the Inquisition on Wednesday, November 25, 1615, it was decided to have Galileo's *Letters on Sunspots* examined. In December 1615, against the advice of his friends and with the permission of the Grand Duke, Galileo went to Rome to protect his reputation, to defend himself against his critics and to promote Copernicanism. Galileo had a high estimation of his argumentative gifts but as Santillana has so justly remarked, it was Galileo's fate through life to create excitement and consensus around him even when it had little to do with real understanding: his extraordinary literary gifts, his brilliant repartee and eloquence and charm as a speaker, in a world which valued such skills, implied an acceptance that was not real. "As he talked reason to his hearers, he believed, he forever wanted to believe, that they were following the course of his thought, and he spent himself unsparingly in explaining and

[27] Finocchiaro, *Galileo Affair*, 83.

persuading. They applauded; but, when time came, this success showed again and again as fool's gold in his hands."[28] Galileo's difficulty was that he was trying to campaign for his views without really appearing to do such and with people he found hard to reach, as he stated in a January 23, 1616, letter to Curzio Picchena (1553–1626), the Tuscan state secretary after Vinta.

> My business is far more difficult and takes much longer, owing to outward circumstances, than the nature of it would acquire, because I cannot speak openly with those persons with whom I have to negotiate, partly to avoid causing a prejudice to any of my friends, partly because they cannot communicate anything to me without running the risk of grave censure. And so I am compelled, with much pain and caution, to seek out third persons who, without even knowing my purpose, may serve as mediators with the principals, so that I can set forth, incidentally as it were, and at their request, the particulars of my case. I have also to set down some points in writing, and to arrange that they should come privately into the hands of those I want to read them, for I find in many quarters that people are more ready to yield to dead writing than to live speech, for the former allows them to agree or dissent without blushing and, finally, to yield to the arguments since in such discussions we have no witnesses but ourselves. This is not done so easily when we have to change our mind in public.[29]

Galileo seemed to be at his best in these small-group meetings at the homes of various Roman hosts. While he won over some, he offended others; and it was to end badly for him. As Monsignor Querenghi, Galileo's old friend from Padua at whose house the literati of Padua met after the death of Pinelli, wrote of him in a letter (January 20, 1616) to Alessandro Cardinal d'Este:

> We have here Mr. Galileo, who frequently in meetings of men with curiousity, attracts the attention of many with regard to the opinion of Copernicus which he holds to be true.... He talks frequently with fifteen or twenty guests who argue with him now in one house, now

[28] Santillana, *Crime of Galileo*, 115–16.
[29] Shea and Artigas, *Galileo in Rome*, 78–79.

in another. But he is so well fortified that he laughs them off; and although people are left unpersuaded because of the novelty of his opinion, still he shows up as worthless the majority of the arguments with which his opponents try to defeat him. Monday in particular, in the home of Federico Ghisilieri, he was especially effective. What I enjoyed most was that before he would answer the arguments of his opponents, he would amplify them and strengthen them with new grounds which made them appear invincible, so that, when he proceeded to demolish them, he made his opponents look all the more ridiculous.[30]

Galileo stayed at the Tuscan embassy, the Villa Medici, for almost six months, much to the discomfort of the Tuscan ambassador, Pietro Guicciardini (1560–1626), who was unsympathetic to him and had opposed his coming to Rome. In a letter of March 4, 1616, to the Grand Duke, a day before the publication of the decree of the Index condemning Copernicus, he described Galileo's behavior:

Galileo has relied more on his own counsel than on that of his friends. The Lord Cardinal del Monte and myself, and also several cardinals from the Holy Office, had tried to persuade him to be quiet and not go on irritating this issue. If he wanted to hold this Copernican opinion, he was told, let him hold it quietly and not spend so much effort in trying to have others share it. Everyone fears that his coming here may be very prejudicial and that, instead of justifying himself and succeeding, he may end up with an affront.... He is all afire on his opinions, and puts great passion in them, and not enough strength and prudence in controlling it; so that the Roman climate is getting very dangerous for him, and especially in this century; for the present Pope, who abhors the liberal arts and his kind of mind, cannot stand these novelties and subtleties; and everyone here tries to adjust his mind and his nature to that of the ruler. Even those who understand something, and are of curious mind, if they are wise, try to show themselves quite to the contrary, in order not to fall under suspicion and get into trouble themselves. Galileo has monks and others who hate him and persecute him, and, as I said, he is not at all in a good position for a place like this, and he might get himself and others into serious trouble.... To involve the Grand Ducal House in these

[30] Langford, *Galileo, Science and the Church*, 80.

embarrassments and risks, without serious motive, is an affair from which there can come no profit but only great damage. I do not see why it should be done, the more so when this happens only to satisfy Galileo. He is passionately involved in this quarrel, as if it were his own business, and he does not see and sense what it would comport; so that he will be snared in it, and will get himself into danger, together with anyone who seconds him.... For he is vehement and is all fixed and impassioned in this affair, so that it is impossible, if you have him around, to escape from his hands.[31]

As far we can see, at no point in his long stay in Rome did he visit or meet with any of the Jesuits who had been his strongest supporters. Certainly, they had reasons to be wary of him, and he was not unaware that his visit and way of proceeding were not according to their wishes. He dedicated his *Discourse on the Tides*, which contained his most explicit Copernican views and his strongest argument for them, to Alessandro Cardinal Orsini, a newly created cardinal, only twenty-three, a cousin to the Grand Duke and related to Cesi. In a consistory of the cardinals with Pope Paul V on February 24, 1616, Orsini interceded for Galileo, but the pope rebuffed him, telling him that he had already turned over the case to the Inquisition. As Guicciardini wrote in his letter to the Grand Duke of March 4: "As he felt people cold towards his intention, after having wearied several cardinals, he threw himself on the favor of Cardinal Orsini, and extracted to that purpose a warm recommendation from Your Highness. The Cardinal, then last Wednesday in Consistory, I do not know with what circumspection and prudence, spoke to the Pope on behalf of the said Galileo. The Pope told him it would be well if he persuaded him [Galileo] to give up that opinion. Thereupon Orsini replied something, urging the cause, and the Pope cut him short and told him he would refer the business to the Holy Office."[32]

Early in 1616, while the Galileo case was being mulled over in Rome, a member of the Congregation of the Index, Bonifacio Cardinal Caetani (1568–1617), who was also Prince Cesi's uncle, called upon one of the most original and striking figures of the time for

[31] Santillana, *Crime of Galileo*, 116–17, 119.
[32] Ibid., 119.

advice: the Dominican Tommaso Campanella. Caetani, whose family was from the old papal aristocracy, the Dukes of Sermoneta and of the family of Pope Boniface VIII, with many cardinals in its train, had been made a cardinal by Paul V in 1606. He was also a highly respected, open-minded individual and secure in Paul V's confidence—and he had to be because the man he called upon could not have been in a more precarious position: an odor of heresy and sedition clung about Campanella, and he had already spent some sixteen years in Neapolitan prisons as the instigator of a rebellion against Spanish rule.

Tommaso Campanella (1568–1639), born of poor parents and joining the Dominicans at thirteen, was a child of Calabria, the land of Pythagoras and Joachim of Fiore, and he shared their visionary mysticism. His extraordinary intelligence, universal interests, insatiable curiosity, amazing eloquence, and unyielding thirst for a new and better world were to make his life a perpetual trial to himself and would earn him decades of imprisonment, often under the most horrendous conditions, including torture. He became fiercely opposed to Aristotle despite the strongly Aristotelian orientation of his order, and in 1589 he began the first of his great peregrinations when he fled from a small, isolated Dominican convent in Calabria for Naples, one of the largest cities of Europe. The highest circles of its intellectual life centered around the della Porta brothers, Giovanni Battista and Giovanni Vincenzo, and the scholars protected by the prince of Conca, participating in the Academy of the Awakened (*Accademia deli Svegliati*), which was closed by the Spanish in 1593.

In 1592 Campanella was imprisoned in a Dominican convent in Naples, and on his release instead of returning to his native province as he had been instructed, he fled north to Rome, to Florence, Bologna (where the Inquisition confiscated all his manuscripts), and eventually to Padua (in 1593), where he enrolled as a student for a few months and met Galileo. Early in 1594 he was arrested by the Inquisition, and after periods of imprisonment and his first experience of torture, he ended up back in Calabria. Believing himself a prophet and expecting a transformation of the world in 1600 (in which a great conjunction of the stars would occur), he became involved in an apocalyptic conspiracy against Spain that led in 1599 to his arrest and imprisonment in Naples, where he was extensively tortured. To save his life Campanella feigned madness, but to test his madness, he was put to the worst trial yet: *la veglia* or "the awakener", where he was suspended over a set of wooden spikes, able only by the use of his arms and shoulder muscles to keep his body from being gashed by them. After forty hours, barely alive but his simulation of madness undetected, he was declared insane and given life imprisonment. He would remain imprisoned in Naples until 1626, when he would go to Rome and become intimately involved with the affairs of Pope Urban VIII. But, despite the dreadful conditions of his imprisonment and blessed with an extraordinary memory since his access to books was limited, he managed to write quite a bit including his impressive *Apologia pro Galileo*, his response to Cardinal Caetani's request.

Campanella greatly admired Galileo. After he read the *Starry Messenger* in January 1611, he wrote a Latin letter to Galileo fulsome in its praise but which also revealed its share of perplexity and hesitation. While he thought Copernicanism probable, not proven, he wished to defend Galileo's right, and the right of all Catholics, to intellectual freedom. He composed the *Apologia pro Galileo* in early 1616 but was not finished until mid-March, too late to help, and it was only published in 1622, in Protestant Frankfurt. The defense was a theological not a scientific work, answering the question of whether Galileo's heliocentric theory was agreeable to the Scriptures, the Church Fathers, and Catholic theology. After a very impressive summary and critique of the evidence from both sides, after pointing out that its judges should be experts both in theology and science, after pointing out that the world is the book of God and that it is to the glory of Christianity that it permitted mankind to discover new sciences and renew old ones, after pointing out that Aristotle did not possess the Christian faith, he stated that Galileo's theory was indeed compatible with the Catholic understanding of Scripture and that to prohibit his investigations and suppress his writings would make a mockery of the Scriptures, damage the Church among the heretics (who would embrace Copernicanism all the more since we reject it), and leave the impression that Catholicism detested great minds.

> Anyone who forbids Christians to study philosophy and the sciences also forbids them to be Christians. Also only the Christian covenant recommends all the sciences to its members, because it has no fear of being shown to be false.... Every human society or law which forbids

its followers to study the natural world should be held in suspicion of being false. For since one truth does not contradict another, as was stated at the Lateran Council under Leo X and elsewhere, and since the book of wisdom by God the creator does not contradict the book of wisdom by God the revealer, anyone who fears contradiction by the facts of nature is full of bad faith.[33]

On February 19, two propositions ascribed to Galileo though not found in any of his writings (but which are found in Caccini's denunciation) were sent to the eleven theological consultors of the Holy Office. The first proposition was that the sun was the center of the universe and immobile; and the second proposition was that the earth was not at the center of the universe and moves as a whole and in a daily motion. On February 23, 1616, after only four days deliberation, they made their unanimous decision on the points, and on February 24 their report was received by the Inquisition. It called the first proposition "foolish and absurd in philosophy, and formally heretical since it explicitly contradicts in many places the sense of Holy Scripture, according to the literal meaning of the words and according to the common interpretation and understanding of the Holy Fathers and the doctors of theology". For the second proposition it judged that "it receives the same judgment in philosophy and that in regard to theological truth it is at least erroneous."[34] Of the eleven theologians, at least five were Dominicans, including Thomas de Lemos, the Spanish theologian famous for his efforts in the *De auxillis* controversy, and one Jesuit, Benedetto Giustiniani, who earlier had allowed for the possibility of Copernican ideas in reference to the philosopher Patrizi.

It seems amazing to us that so momentous a decision should be decided so quickly, so seemingly cavalierly and by such a simple bureaucratic exercise. As Olaf Pedersen has pointed out, a few centuries earlier, such a problem would have been referred to the great universities and the disputations of the larger community of scholars, to the serious weighing of arguments pro and con by experts, and the sifting of evidence, scientific and theological. The Reformation and

[33] Campanella, *Defence of Galileo*, 54, 68–69.
[34] Finocchiaro, *Galileo Affair*, 146.

the Counter-Reformation had changed all that, and while the universities still had some clout, the bureaucratic apparatus of the Holy See had gained the upper hand.[35]

On February 25, at the Thursday meeting of the Inquisition, having heard the report of the theologians against the propositions of Galileo, Pope Paul V ordered Bellarmine to meet with Galileo and warn him to abandon his Copernican opinions, and if he refused, to have the Commissary of the Inquisition in the presence of a notary and witnesses give him a precept that he completely abstain from teaching or defending this doctrine, or discussing it, and if he did not agree, to threaten imprisonment (arrest). In the minutes of the Inquisition for March 3, 1616, it was reported that Bellarmine had given the report that "the mathematician Galileo Galilei had acquiesced when warned of the order",[36] and no special precept is mentioned. In the files of the Holy Office later on, however, was to be found an unsigned document (though clearly in the same handwriting of the rest of the file) from February 26, which stated that Galileo was given the precept "to abandon completely" the condemned opinion "and henceforth not to hold, teach, or defend it in any way whatever, either orally or in writing".[37] This document went beyond what Pope Paul V had asked for and what the minutes stated; it also went beyond what was stated in the certificate that Bellarmine had given to Galileo later, on May 26, to allay rumors of his condemnation which stated that these opinions only could not "be defended or held".[38] When it was brought up in 1633, Galileo had no memory of it, but it would play an important part in the 1633 trial.

Many theories have been proposed to answer this discrepancy, including ideas of the later planting of a forgery to trip Galileo up, and it is unlikely that we will ever know with certainty, but it seems more likely that Galileo demurred in some way and the stronger precept was given him.[39] What is also mysterious, but perhaps easier to

[35] Olaf Pedersen, "Galileo and the Council of Trent: Galileo Affair Revisited", *Journal of the History of Astronomy* 14 (1983): 24.

[36] Finocchiaro, *Galileo Affair*, 148.

[37] Ibid., 147–48.

[38] Ibid., 153

[39] Thomas F. Mayer, "The Roman Inquisitions Precept to Galileo (1616)", *British Journal for the History of Science* 43 (2010): 327–51.

explain, is why, after the clear condemnation of Galileo's two points, including one as "formally heretical", much more was not done with Galileo, and why when the decree of the Index was published on March 5, 1616, it was so much tamer and he was unnamed.

The Copernican doctrine was declared to be not heretical, the highest form of ecclesiastical censure, but "altogether contrary to Scripture", a somewhat lesser and more ambiguous censure; and Galileo was not named at all despite the fact that his *Letters on Sunspots* supported Copernicanism and these decisions to a great extent were in response to Galileo's activities. Why? That Galileo had powerful patrons in the Medici and was a celebrated scientist, an international superstar one might say nowadays, would certainly explain much, but not all. A secondary explanation, and one that we have evidence for, is that two powerful cardinals, Barberini and Caetani, objected to the stricter judgment. This is referred to in an account from 1633 by Giovanni Francesco Buonamici (1592–1669), who served as a high-level diplomatic secretary at various courts, including as secretary to the Tuscan ambassador to the court of Pope Paul V, and was later a close friend of Galileo (his wife's sister married Vincenzio, Galileo's son). After discussing Galileo's amazing discoveries with the telescope, he refers to the opposition he encountered.

These and other sensible demonstrations, which Galileo with the benefit of the telescope discovered in the heavens before anyone else, excited envy in many persons; being envious for his glory and unable to contradict the manifest truth of the discoveries made in the heavens, they started to persecute him; this was especially true for some Dominican friars who took the road of the Inquisition and Holy Office in Rome, complaining that he attributed stability to the sun and mobility to the earth against the words of Sacred Scripture. Thus Paul V, instigated by the same friars, would have declared this Copernican system erroneous and heretical, insofar as it is contrary to the teaching of Scripture in various places and especially in Joshua, had it not been for the opposition and defense of Lord Cardinal Maffeo Barberini (today Pope Urban VIII) and of Lord Cardinal Bonifazio Caetani. However, these cardinals argued, first, that Nicolaus Copernicus could not be declared heretical for a purely natural doctrine, without eliciting the laughter of the heretics, who do not accept the reform of the calendar of which he was the principal master; second, that it did not seem prudent to assert on the authority of Sacred Scripture in purely natural subjects that something is true which (with the passage of time and by means of sensible demonstrations) could be discovered to be false, for even in subjects concerning Faith (which is the principal if not only purpose of Sacred Scripture) it is frequently necessary to understand that it speaks in accordance with our abilities, otherwise if one wanted to abide by the pure sound of the words one would end up in errors and impieties (such as that God has hands, feet, emotions, etc.). Thus, these cardinals dissuaded Paul V from the ruling that the said friars had come close to extracting from him.[40]

It would not be surprising from what we have already seen that Cardinal Caetani opposed the more severe condemnation, and it is most likely to him that were owed the very moderate corrections that were eventually made to the text of Copernicus' great work. That Cardinal Barberini opposed the condemnation of Copernicus in 1616 is confirmed by a number of other sources. In a letter from Castelli to Galileo of March 16, 1630, he recounted that he heard from Prince Cesi that Campanella had recently informed Pope Urban that he had nearly converted some German Protestants, but they had recoiled because of the decree against Copernicus. The pope reportedly answered that the condemnation of Copernicus was not the Church's intention, and if it had been left to him, the decree would never have happened. Also, in a conversation in 1624 in Rome between Galileo and Friedrich Cardinal von Hohenzollern-Sigmaringen (1582–1625), prince-bishop of Osnasbrück, Galileo learned that Pope Urban had told the cardinal that the Church had never condemned Copernicanism as heretical, just as rash. That Pope Paul V had originally wanted to condemn it as heretical (with Bellarmine's support) was also what Guiccardini the Florentine ambassador reported to the Grand Duke in a letter of March 4, 1616.[41]

The decree from the Congregation of the Index announcing what corrections were to be made to Copernicus' *De revolutionibus* came out in May 1620—and it was the only time that the Congregation of the Index ever announced specific corrections—with about a dozen

[40] Maurice A. Finocchiaro, *Retrying Galileo, 1633–1992* (Berkeley: University of California Press, 2005), 34.

[41] Jules Speller, *Galileo's Inquisition Trial Revisited* (Frankfurt: Peter Lang, 2008), 77–79.

corrections, a remarkably small number considering it is a rather large work of some 400 pages and 146 diagrams. In the report made to the Congregation of the Index on how to proceed with the book's corrections, it gives its rationale: first, that since astronomy is important to the Church because of its connections with the calendar, the book must be preserved for its observations and for the restoration of astronomy; second, that the removal of heliocentrism would not be so much a correction as a destruction of the system; third, that it would be possible to find a solution that would not compromise Scripture by changing certain passages. "If certain of Copernicus' passages on the motion of the earth are not hypothetical, make them hypothetical; then they will not be against either the truth or the Holy Writ. On the contrary, in a certain sense, they will be in agreement with them, on account of the false nature of suppositions, which the study of astronomy is accustomed to use as its special right."[42] It seems that the decree was barely enforced, since of the copies of the book from the 1543 (Nuremberg) and 1566 (Basel) editions now in existence, over 90 percent are not censored, with about two-thirds of the Italian copies censored and very few elsewhere, and none in Catholic Spain, whose own Index explicitly permitted it; and even in Italy the quality of censorship varied greatly, with some copies completely having text inked out or covered and others, as in Galileo's personal copy, where the deleted text is quite visible.[43] Among those copies that are uncorrected is that from the Jesuit Roman College.[44]

Foscarini died in June 1616, with his rejoinder to Bellarmine unfinished and his book prohibited. His publisher in Naples closed up shop but was arrested, found guilty by the Inquisition, and fined a hundred ducats. Galileo remained in Rome, and a week after the decree of the Index he had an audience with Pope Paul V for three-quarters of an hour. In a letter to Curzio Picchena he wrote: "I pointed out to His Holiness the maliciousness of my persecutors and some of their false calumnies, and here he answered that he was aware of my integrity

[42] Owen Gingerich, "The Censorship of Copernicus' *De revolutionibus*", *Journal of the American Scientific Affiliation* (March 1981): 58–60; Owen Gingerich, *The Book That Nobody Read: Chasing the Revolutions of Nicolaus Copernicus* (New York: Walker & Company, 2004), 146.

[43] Ibid.

[44] John L. Heilbron, *The Sun in the Church* (Cambridge, Mass.: Harvard University Press, 1999), 18.

and sincerity. Finally, since I appeared somewhat insecure because of the thought that I would be persecuted by their implacable malice, he consoled me by saying that I could live with my mind at peace, for I was so regarded by His Holiness and the whole Congregation that they would not easily listen to the slanderers, and that I could feel safe as long as he lived."[45] Because of the many rumors that Galileo was hearing about from his friends, rumors that he had been brought to the Inquisition, forced to abjure heresy and do penance, in late May Galileo asked Bellarmine for a declaration of what had really happened. He received a certificate from Bellarmine saying that he had not been called before the Inquisition to be abjured of heresy or had received any penance, only that he had been notified of the decision of the pope and the Index that Copernicus' views were contrary to Scripture and could not be defended or held. Galileo received letters from Curzio Picchena on March 20 and very clearly on May 23 telling him to leave Rome and return home. He finally left Rome in early June, furnished with testimonies from Cardinals del Monte and Orsini of his good reputation. While he tried to put a good face on it, it could not be doubted that he had failed. But there was still the future, and Galileo was an optimistic, self-confident, and tenacious man.

The question has to be asked if Galileo himself was the cause of the 1616 condemnation of Copernicus: that his intemperate and zealous advocacy, even to the point of going to Rome itself and challenging the powers that be, made a decision against him inevitable. There are those, and not merely pious defenders of the status quo, who, looking at the comments of so many of Galileo's friends in Rome and particularly of those associated with the Jesuit scientists at the Roman College that all was well for the time being, and knowing of Bellarmine's view of Copernicanism as being heretical if promoted as a true description of the universe, believe that a quiet campaign would have been more successful in the long-term.[46] While it is true that the tactics of opponents were shameful and hurtful to Galileo, and that his intentions were good (for he seemed to fear that a condemnation was imminent despite what his friends said), the fact that Galileo

[45] Finicchiaro, *Galileo Affair*, 152.

[46] Mordechai Feingold, "The Grounds for Conflict: Grienberger, Grassi, Galileo, and Posterity", in *The New Science and Jesuit Science: Seventeenth-Century Perspectives*, ed. Mordechai Feingold (Dordrecht: Kluwer, 2003), 121–57.

lacked decisive demonstrations that would have proved his case and forced the issue left the authorities little choice. (Decisive physical evidence would not be available for quite some time, beginning with James Bradley's observance of stellar aberration in 1729, to Bessel's sighting of the stellar parallax in 1838, and to the demonstration of Foucault's pendulum in 1851). Nor can there be any doubt that Galileo could unnecessarily antagonize people, and he already had many enemies in Rome. While it is true that Foscarini's book had brought the question up publicly, it seems quite possible that something much less damning could have happened that left the Church freer in reference to the New Science. For once a decision was made, it would be very hard for such an organization as the Catholic Church to go back on it, as the succeeding centuries show.

In June 1614, Christopher Grienberger, who had succeeded Clavius as professor of mathematics at the Roman College and head of the Academy of Mathematics there and who had always been friendly toward and supportive of Galileo, far more than Clavius himself, invited Giovanni Bardi, a student at the Roman College and Galileo's friend, to give a lecture and demonstration on floating bodies at the college, with Grienberger doing the demonstration himself. Such exercises were part and parcel of the attempts of the Jesuit mathematicians to raise the status of their discipline and show the intrinsic worthiness of their efforts. Three years earlier Galileo himself had given a demonstration on floating bodies in opposition to traditional natural philosophers, which led to his 1612 *Discourse on Floating Bodies*, his first published work on physics. After the lecture and demonstration, Bardi wrote Galileo about Grienberger's comments to him.

> Fr. Grienberger told me that if the topic had not been treated by Aristotle (with whom, by order of the General, the Jesuits cannot disagree in any way but are rather obliged always to defend), he would have spoken more positively about the experiments because he was very favorably impressed by them. He also told me that he was not surprised that I disagreed with Aristotle, because Aristotle is clearly also wrong in regard to what you [i.e., Galileo] once told me about two weights falling faster or slower.[47]

---

[47] Blackwell, *Galileo, Bellarmine, and the Bible*, 136.

Grienberger was an extremely modest man who had no trouble in assisting others and giving them credit for his work; and it seems that this lecture—which would give public legitimation to Galileo's own anti-Aristotelian views—was most likely authored not by Bardi but by Grienberger himself. This was not just due to his personal modesty, or even to his conformity to the Jesuit ideal of self-abnegation, to efface whatever could to lead to vainglory, but because it would be much easier for him, perhaps even necessary, to hide behind a non-Jesuit where the result would lead to opposition to Aristotle whom the Jesuit order was committed to defending.[48] Grienberger had frequently and significantly intervened in Galileo's favor a number of times, supporting him over the Jesuit Scheiner on sunspots, for example, but always privately or behind the scenes. All of these actions were part of a much larger "rear-guard campaign" led by Grienberger to discredit Aristotelian natural philosophy and elevate the mathematical sciences.[49] Bardi's comments about the order of the general refer to two letters of Claudio Acquaviva (1543–1615), one from 1611 which expressed a need for uniformity in broad terms and did not specifically mention philosophy, and another, much more specific and insistent one, from 1613, which enjoined the Jesuits to return to "solid and uniform doctrine" in their teaching and writing on theology and philosophy, following the theology of Saint Thomas and the philosophy of Aristotle.

From their very beginning the Jesuits had been an order of scholars (we sometimes forget that the original group were all masters or graduate students from the great University of Paris), and while they stressed uniformity of thought and vision, they understood that a certain degree of liberty was also necessary to achieve their goals. Both the Jesuit *Constitutions* and the *Ratio Studiorum* of 1599, which organized Jesuits' studies, explicitly warned them to shun novelty and follow Aristotle and Saint Thomas in their philosophical and theological studies; but attempts in the early decades to control the content in a much more detailed way generally failed.[50] However, in

---

[48] Michael John Gorman, "Mathematics and Modesty in the Society of Jesus: The Problems of Christoph Grienberger", in *The New Science and Jesuit Science: Seventeenth-Century Perspectives*, ed. Mordechai Feingold (Dordrecht: Kluwer, 2003), 25–26.

[49] Feingold, "Grounds for Conflict", 121–57.

[50] Marcus Hellyer, *Catholic Physics: Jesuit Natural Philosophy in Early Modern Germany* (Notre Dame, Ind.: University of Notre Dame Press, 2005), 13–30.

the early seventeenth century, in the later years of the generalate of Acquaviva, the balance shifted away from liberty toward uniformity. Claudio Acquaviva, general of the Jesuits from 1581 until his death on January 31, 1615, and often referred to as their second founder, weathered many storms (the revolt of the Spanish Jesuits supported by Philip II, the hostility of the Franciscan Pope Sixtus V and not a few monarchs, the *De auxiliis* controversy), and led the Jesuits in a period of remarkable growth, leaving them nearly three times as large as when he entered, from approximately five thousand to about thirteen thousand members. While there were many intrinsic advantages to uniformity, practical and pedagogical, as the Jesuit order grew ever larger and their schools became ever more numerous, those advantages became even greater and more desirable. But there also was another reason for the more aggressive promotion of "solid and uniform doctrine", and it was connected to the *De auxiliis* controversy, which we saw in the first chapter.

In 1607 Paul V had ended the decades of controversy over grace and free will which was the *De auxiliis* controversy by sending the participants away to await a decision that never came. He also demanded that both sides cease from their attacks on each other under pain of severe punishment. But it was a precarious peace, and since it touched upon some of the most important elements of the faith, it was hardly likely ever to go away, as the rise of Jansenism a few decades later shows. A book on the topic of grace by the Belgian Jesuit Leonard Lessius (1554–1623), finished and approved in 1608, was allowed to be published in Antwerp in 1610, though only after another on the same topic by the Dominican Diego Alvarez (1550–1635) had been published. But Lessius' book prompted the Spanish king Philip III to push again for a definite settlement and the General Chapter of the Dominicans in 1612 petitioned the pope for the same. On December 1, 1611, the Inquisition published a decree prohibiting publication of further works on the doctrine of grace without its special permission. The Jesuits, while successful in their earlier endeavor not to be condemned, had been left slightly bruised and somewhat cautious, and the fear of reigniting the controversy was so great as to cause the two letters of Acquaviva. The 1613 letter, which was to be read annually at faculty meetings at the beginning of each academic year, went far beyond any other rulings and raised

Aristotle to a level not seen before.[51] It also allowed for the removal of teachers from their positions, even in the middle of their courses if necessary, and their reassignment to other duties if they failed to follow these prescriptions. Christoforo Borri (1583–1632) was removed from his teaching position in Milan in 1614 after complaints were received that he was teaching novelties, including the doctrine of the fluidity of the heavens. Things would only become worse after Acquaviva's death in January 1615 and the accession of Muzio Vitelleschi (1563–1645) as the new general, and after the 1616 decree of the Index condemning Copernicanism.

Giuseppe Biancani (1566–1624), born in Bologna and entering the Jesuits in 1592, studied at the Roman College and then from 1596 to 1599 at the Jesuit College in Padua, and from the early 1600s for the rest of his career he taught mathematics at the Jesuit College at Parma, and even at the university there. When he studied at Padua he became acquainted with Galileo, who was teaching at the university, and developed a lasting admiration for him. In a letter of June 14, 1611, he wrote of Galileo: "I love and admire Galileo, not only for his singular learning and inventions, but also for the lasting friendship I struck up with him in Padua, where I was overcome by his courtesy and affection, that bound me to him. No one, I believe, has spread and defended his findings, both in public and in private, more than I have."[52] That being said, he had no qualms about disagreeing with Galileo, as he seemingly did in 1611 over the existence of mountains on the moon and later in 1613 when he supported Scheiner in the controversy over sunspots. In 1614 he finished *Aristotelis loca mathematica*, a critical analysis of the mathematical parts of Aristotle and its uses in the sciences. The Jesuit censor Giovanni Camerota (1559–1644) did not think it proper or useful for the books of a Jesuit ("our members") to contain the ideas of Galileo, particularly when they were contrary to Aristotle, nor say that Aristotle's view was wrong or his arguments defective; nor should they praise heretics such as Maestlin, Tycho, and Kepler,

---

[51] Ibid., 33.

[52] Francesco Paolo de Ceglia, "Additio Illa Non Videntur Edenda: Giuseppe Biancani, Reader of Galileo in an Unedited Censored Text", in Feingold, *New Science and Jesuit Science*, 159.

as Biancani had done. He also objected to Biancani's support for Tycho's opinion that comets are above the moon (and so the heavens are changeable), the heavens are fluid matter, and the planets move through by their own motive force; and he objected most especially to his support for Galileo's views on floating bodies against Aristotle. In the end the book was published in 1615, but Biancani had to tone down its critical attitude toward Aristotle, speak neutrally, and report without commitment positions seemingly contrary to official doctrine, and refer to Galileo's treatise on floating bodies without comment. His general work on astronomy completed in 1615, *Sphaera mundi, seu cosmographia*, where he presented details of the work of Copernicus, Brahe, Kepler, and Galileo, was also heavily criticized, and despite the internal pleas of Grienberger for greater of freedom of thought in this matter.

A new *cosmographia* seems to be necessary because the old one has been changed a great deal in our day and many embellishments have been added to it. But the question has been raised as to whether it is proper for us Jesuits to do this. It seems to me that the time has now come for a greater degree of freedom of thought to be given to both mathematicians and philosophers on this matter [constitution of the heavens], for the liquidity and corruptibility of the heavens are not absolutely contrary to theology or to philosophy and even much less to mathematics.... It seems that he [Biancani] has not exercised his talents sufficiently in writing the *Cosmographia*. But I am quite willing to excuse him about this. For up to now his hands have been tied, as have ours. Thus he dealt with most topics in a way which is not adequate when he was not allowed to think freely about what is required.[53]

In the end it could only be published in 1620 and with significant censoring. The language was softened and the new ideas were still reported but in a neutral way and in conformity to the decree of 1616 from the Index.[54] At the very time when science was moving

[53] Feingold, "Grounds for Conflict", 136.

[54] Blackwell, *Galileo, Bellarmine, and the Bible*; and J.J. O'Connor and E.F. Robertson, "Giuseppe Biancani", School of Mathematics and Statistics at the University of St. Andrews, Scotland, July 2012, http://www-history.mcs.st-andrews.ac.uk/Biographies/Biancani.html.

away from Aristotle, Jesuit scientists, already governed by their vows of obedience and by strict censorship in their publications and public teaching, found themselves obliged to defend him all the more. And the condemnation of Copernicus in 1616 only made this even more difficult. Their campaign for the New Science and their attempts to elevate the status of the mathematical sciences could only then be covert, by the use of subterfuges and dissimulations. What was left was a studied ambiguity which leaves much unsaid, as we see, for example, in Biancani's discussion of heliocentrism from his *Sphaera mundi*.

This ancient [heliocentric] belief was brought back to life again in the past century by Nicolaus Copernicus, a man of sharp mind and a great restorer of the science of astronomy. He even defended it against the argument of other men, so that today some mathematicians of high repute, as for example Johannes Kepler, William Gilbert (author of "De magnetica philosophia"), and others, support the same unfortunate view. Other mathematicians reject this view as absurd. Copernicus actually added that not only is the earth moved in the ecliptic, but together with the earth, the water and the air and all the "interlunary" sphere, in just the same proportion as the earth moves, by which hypothesis not only do Copernicus and his followers seek to save appearances but also believe they easily escape the counter-arguments of their opponents. For all that, however, this opinion is false and must be rejected (even if supported by better proofs and arguments), as is manifest by reasons given formerly and by many opinions of authorities; all the more because it has been forbidden by the ecclesiastical authorities as contrary to Holy Writ.[55]

After 1616 Jesuit scientists had to be careful, but whether Jesuit science "died on the vine, just as the first blossoms had appeared", or it was "the beginning of the end for classical Jesuit science" or the "onset of intellectual rigor mortis" for the Jesuits is not so clear.[56] That the order could continue to produce scientists, including men of such first rank as Roger Boscovich (1711–1787), is a sign of some significant degree of vitality, and there are a number of scholars who will point to the fact that actual Jesuit practice allowed for a greater

[55] O'Connor and Robertson, "Giuseppe Biancani".
[56] Blackwell, *Galileo, Bellarmine, and the Bible*, 142, 152, 153.

divergence from official standards, the frequent reiteration of the need for uniformity being a good indicator that much more diversity occurred than was officially presented.[57] The Jesuits did keep abreast of contemporary scientific research and even contributed to it. While the restrictions and censorship could be irksome and delay the publication of new works, it doesn't seem, of itself, to have destroyed Jesuit science, and there were always ways around official restrictions. As Heilbron writes: "Even the Jesuits were teaching heliocentrism before the end of the seventeenth century, using the convenient fiction that it was a convenient fiction. Those willing to call a theory a hypothesis could publish any astronomy they wanted."[58]

Up to 1616, the Jesuits had been the greatest support to Galileo, and time and again he had turned to them as the beacons of scientific excellence and integrity in all his difficulties. Their confirmation of his telescope findings had been a significant element in his initial acceptance, and during the attempts of the Pigeon League to have him condemned, he had appealed to them repeatedly. After 1616 that would start to change, until in the mind of Galileo they became his greatest enemies, and he blamed them for his troubles because he had challenged their intellectual dominance of the Catholic world. Thus he could write, now under perpetual house arrest and unable to publish, in a letter of March 16, 1635, to the French savant Nicolas-Claude Fabri de Peiresc (1580–1637): "Having discovered many fallacies in the philosophies commonly taught in the schools now for many centuries and having communicated some of them and also published some, I have aroused such animus in the minds of those who want themselves alone to be known as learned that, because they are very crafty and powerful, they have known how and been able to grasp the means to suppress what I found and published and to impede my publication of what remains with me."[59] In a July 25, 1634, letter to Diodati he spoke of "very powerful enemies" who had finally chosen to reveal themselves, "for about two months ago a good friend of mine was in Rome talking to Father Christopher Grienberger, a Jesuit and a mathematician at that College; when they came

---

[57] Hellyer, *Catholic Physics*.
[58] Heilbron, *Sun in the Church*, 22.
[59] Westfall, *Essays on the Trial of Galileo*, 50.

to discussing what happened to me, the Jesuit said these exact words 'If Galileo had been able to retain the affection of the Fathers of this College, he would be living in worldly glory; none of this his misfortunes would have happened; and he could have written at will on any subject, I say even of the earth's motion, etc.' So Your Lordship sees that it is not this or that opinion which has provoked and provokes the war against me, but to have fallen out of favor with the Jesuits."[60] This was unfair. In September 1616, Grienberger and Guldin had spoken to Prince Cesi of their affection for Galileo and their disgust at the recent proceedings.[61] And, as we have seen, many Jesuits continued to speak well of him even at a cost to themselves. Despite what Galileo and many others have thought, for the Jesuits soon became the villains of the Galileo Affair all across Europe, there seems to have been no Jesuit conspiracy in 1632–1633 to bring Galileo down.

Great organizations have great strengths, but they also have their limitations. Knowing their dominance of Catholic scholarship and education, knowing their prestige among the Catholic elites, Galileo fully appreciated what the support of the Jesuits could mean for him. What Galileo could not fully understand and accept were the limitations that came with a large organization: that as a small part of a vast religious order, and under vows of obedience, the Jesuit scientists were not free agents to do as they pleased. He could not understand that in any group personal interests and ambitions must be subsumed to achieve the larger goals. And this despite the fact that Grienberger, through common friends and in personal letters, both implicitly and even explicitly in one case, told him that he did not have this freedom.[62] Grienberger believed that these goals would best be achieved by caution, worldly prudence, and self-effacement. Galileo was never a "team player" as the saying goes, and, after his telescopic achievements, he had become a "rock star" with people fawning over him, with no need and no desire to manifest caution, worldly prudence, and self-effacement. His alienation from the Jesuits, which one must say was primarily of his own making, would lose him valuable allies in the future. The great mathematician Torricelli,

---

[60] Finocchiaro, *Retrying Galileo*, 58.
[61] Feingold, "Grounds for Conflict", 135.
[62] Ibid., 128.

introducing himself to Galileo in a letter of September 11, 1632, mentioned the fact that his good friend Grienberger had said that the *Dialogue* had given him "the keenest delight, and that there were many fine things in it", even if the Copernican theory should have not been extolled but presented "as plausible only, and not true".[63] I think it is more likely to see in those comments of Grienberger's spoken earlier, after the trial of 1633, as Feingold does, not as some indicator of a Jesuit plot against Galileo but as a lament that by alienating the Jesuits as he did, "Galileo effectively eliminated the ability of those most favorable to his ideas quietly to effect changes from within the Catholic Church, which might have averted his own condemnation as well as the collapse of Grienberger's life-long campaign for a freer Jesuit science."[64]

After Galileo's return to Tuscany he continued to work though he was often sick, particularly returning to his work on motion that would later become his groundbreaking *Two New Sciences* (1638). Nor did he give up his astronomical interests and observations. In May 1618 he sent Archduke Leopold of Austria (brother of the Dowager Grand Duchess Christina) two telescopes as well as his *Letters on Sunspots* and *Treatise on Tides* after the archduke had asked for something of his. He rented a villa outside Florence in 1617, one with an excellent view of the city, where he tended to spend most of his time. He had placed his two daughters Virginia and Livia in a Poor Clare (Franciscan) convent nearby at Arcetri in 1613. (At sixteen the daughters would take vows of religion there and join the community permanently, Virginia taking the name Maria Celeste and Livia taking the name of Arcangela.) He would have his son Vincenzo legitimized in 1619, sending him to the University of Pisa, where he gained a doctorate in law in 1628. In May and June 1618 he went on a pilgrimage to one of the great religious shrines of Italy, the Holy House of Loreto, said to be the house of the Holy Family (Jesus, Mary, and Joseph) brought from the Holy Land by angels. He also labored to find a practical and easy way to find longtitude, which would have been invaluable to seamen and merchants (and navies too) and even negotiated with the Spanish about it, though it came to nothing.

[63] Ibid., 153.
[64] Ibid., 154.

This relatively peaceful time was broken by the startling appearance in 1618 of three comets, one after another, over a short span
of time, between August and November, the last of which was particularly impressive beginning in November and disappearing only
in January 1619. Comets of this magnitude had not been seen since
1577. Historically, comets had been seen as harbingers of disaster,
and 1618 would see the beginning of the Thirty Years' War, which
would kill eight million people and leave Germany devastated. While
Galileo was too sick to observe the comets and remained in bed,
others did, including the Jesuit Orazio Grassi (1583–1654), from a
noble family of Savona in the Republic of Genoa and holder of the
chair of mathematics at the Roman College following Grienberger,
from 1617 to 1624 and 1626 to 1632. He gave an astronomical disputation (a public lecture) on them and published in 1619 a ten-page
pamphlet entitled *An Astronomical Disputation on the Three Comets of
1618* "by one of the Fathers of the Roman College", Grassi's name
not being used. Aristotle believed comets were caused by gases rising
from the earth that burned when they rubbed against the crystalline
sphere that held the moon. Following the opinion of Tycho Brahe,
Grassi believed that these comets were beyond the moon, though he
still believed that they were orbiting the earth, which conformed to
Aristotle. There was no criticism of Galileo in Grassi's work, and he
even explicitly credited Galileo for all his previous astronomical discoveries; and, as Sharratt points out, it could even be seen as helping
Galileo indirectly insofar as it showed the reliability and usefulness of
the telescope, which had already revealed so many new things and
might be useful in settling disputes about comets.[65]

A Roman friend of Galileo's, Giovanni Battista Rinuccini, however, wrote him that some people, though not people associated with
the Jesuits, thought that this work overthrew the Copernican system,
and insisted that he present his opinion. Mario Guiducci, a friend
and student of Galileo, gave a lecture on the comets at the Florentine
Academy, publishing it in June 1619 as the *Discourse on the Comets*.
In fact, the bulk of it was written by Galileo, and it was a broadside
against Tycho Brahe, Grassi, and the Roman Jesuits. Galileo believed
that the comets could just be an optical illusion caused by the vapors

---

[65] Michael Sharratt, *Galileo: Decisive Innovator* (Cambridge: Cambridge University Press,
1994), 134.

rising from the earth extending beyond the moon, which the sun then illumined, making them visible to us. In a July 1619 letter of Grienberger we see how greatly the Roman Jesuits were offended by the work:

> As to the affairs of Galileo, I would prefer not to get mixed up in them after he has behaved so badly with the mathematicians of the Roman College, by whom he was treated, in fact, more than once not less well than with sincerity. If even one mention of him had been made in the *Disputatio Romana* or if he had been refuted in any way, I would be less resentful towards him. But since no thought at all was given to him and the whole question turned on the fact that the comets were found much higher than common opinion maintained and use was made here of an hypothesis which up to now it had been licit to admit, I cannot marvel enough as how it could have leapt into Galileo's mind to consider himself under attack, and how he has preferred to suspect temerariously, rather than excuse if perhaps some statement less in conformity or even contrary to his opinions had been made. It has, therefore, been enough for me to give a reading to his academic discourse. And it has been enough for me to know once for all that which he feels in our regard.[66]

Grassi replied with *The Astronomical and Philosophical Balance* (*Libra astronomica ac philosphica*) under the name of Lothario Sarsi (an anagram for his name). It was to this book that Galileo responded, four years later, with his famous *The Assayer* (*Il Saggiatore*) in August 1623. It was a literary tour de force. While a kind of commentary on Grassi's book, it contained much of his view of science, which would become *our* view of science, and included this beautiful passage:

> Philosophy is written in this grand book the universe, which stands continually open to our gaze. But the book cannot be understood unless one first learns to comprehend the language and to read the alphabet in which it is composed. It is written in the language of mathematics, and its characters are triangles, circles, and other geometric figures, without which it is humanly impossible to understand a single word of it; without these, one wanders about in a dark labyrinth.[67]

---

[66] Feingold, "Grounds for Conflict", 141–42.

[67] Stillman Drake, *Discoveries and Opinions of Galileo* (New York: Anchor Books, 1957), 237–38.

But it also was meant, by its polemical and satirical force, to humiliate Grassi and make him ridiculous. And in this it succeeded only too well. Robert Westfall has called it "one of the all time masterpieces of sarcastic invective", which not even a saint could have received without hostility, "and Grassi has not been nominated for sainthood."[68] While a certain combativeness was expected of someone in Galileo's position, and disputes were part and parcel of academic life, Galileo took it to a new level. As the Vatican astronomer (an American Jesuit) Guy Consolmagno has pointed out, while *The Starry Messenger* was filled with the light and joy and wonder of discovery, in *The Assayer*, however, we see, despite the literary brilliance, a querulous old man more interested in scoring debating points than enjoying truth for its own sake. And with its sarcasm and out-and-out lies, it permanently alienated people who might have been best positioned to help him later on: the Jesuit scientists whom he had ridiculed. It may seem odd that Galileo should have done this, but as Consolmagno also points out, he had three strong emotional reasons for wanting not to believe in comets as a distant astronomical object. First, a real, distant comet caused problems for Copernican theory since it showed no retrograde motion. Second, Galileo hated Tycho Brahe and all that he stood for because Tycho's theory of the solar system was the most serious rival to Copernicus. Tycho's theory also got around a great number of scientific problems that Copernicus' theory could not handle, and it was also consistent with the best physics of the day. Third, there was Galileo's own jealousy. He had first used the telescope, but he had never seen the comets since he was sick in bed at the time. In his hurt pride he was forced to declare that these events were not astronomical at all.[69]

Feingold, rightly I think, sees Galileo's objections to the talk, and all that followed it, as rooted in an incapacity to tolerate competition and in a desire to avenge himself on the Jesuits. In Galileo's marginal notes on Grassi's later work, *Ratio ponderum* (1627), we find this: "You cannot help it, Mr. Sarsi, that it was granted to me alone to discover all the new phenomena in the sky and nothing to anybody else. This

---

[68] Westfall, *Essays on the Trial of Galileo*, 51.

[69] Guy Consolmagno, "Precursors of Evil", in *Brother Astronomer: Adventures of a Vatican Scientist* (New York: McGraw-Hill, 2000), 61–79.

is the truth which neither malice nor envy can suppress." And in Galileo's *Dialogue* we see: "The original discoverer and observer of the solar spots (as indeed of all other novelties in the skies) was our Lincean Academician [Galileo]." One could also add that while Grassi did refer to Tycho, it was because he was the expert on the comet of 1577 and there was no explicit support for Tycho's cosmology in the work.[70] Finally, that Galileo's paranoia extended to the Jesuit scientists of the Roman College we have his own words from the *Assayer*: "Sarsi has, therefore, no reason for saying that I am quite guilty of contempt for the dignity of the Collegio Romano. Quite the contrary; if Sarsi's voice does emanate from the Collegio, I have reason to suspect that my doctrine and my reputation have been in bad odor there not merely at present but all along, inasmuch as in this *Balance* none of my thoughts are approved. Nothing is to be read there except opposition, full of accusation and blame, and, if one may believe the rumors, there is in addition to what is written an open boast of power to annihilate everything of mine."[71] While he immediately adds that he does not believe this, the damage has already been done, and the whole tenor of the piece goes against this.

In October 1622 he had finished the book, and copies of the manuscript were sent to other members of the Lincean Academy for comments. The manuscript was approved by the Dominican Niccolò Riccardi, and his official imprimatur ("let it be printed") included in the published volume went far beyond a simple recognition of it not being contrary to the Catholic faith:

> By order of the Most Reverend Father, Master of the Sacred Palace, I have read this work, *The Assayer*; and besides having found here nothing offensive to morality, nor anything which departs from the supernatural truth of our faith, I have remarked in it so many fine considerations pertaining to natural philosophy that I believe our age is to be glorified by future ages not only as the heir of works of past philosophers but as the discoverer of many secrets of nature which they were unable to reveal, thanks to the deep and sound reflections of this author in whose time I count myself fortunate to be born—when

---

[70] Feingold, "Grounds for Conflict", 137–41.

[71] Stillman Drake and C. D. O'Malley, *The Controversy on the Comets of 1618* (Philadelphia: University of Pennsylvania Press, 1960), 179.

the gold of truth is no longer weighed in bulk by the steelyard, but is assayed with so delicate a balance.[72]

While it was being printed in Rome, a most extraordinary and unheralded thing happened—and even Galileo thought it a miraculous conjunction: Galileo's longtime admirer Maffeo Barberini was elected pope. Seizing the opportunity, he dedicated the book to him.

[72] Westfall, *Essays on the Trial of Galileo*, 65.

Chapter Five

Galileo Agonistes: *The Dialogue on the Two Chief World Systems,* The Trial of Galileo and After

In 1621 Pope Paul V died (as did Cardinal Bellarmine), to be succeeded, with Scipione Cardinal Borghese's assistance, by the first Jesuit-trained pope, Alessandro Ludovisi as Pope Gregory XV (r. 1621–1623). While two of Galileo's friends found favor in the new regime, Giovanni Ciampoli (named Secretary for Latin Briefs, and later Secretary for Letters to Princes) and Virginio Cesarini (named Secret Chamberlain to the pope), both Linceans since 1618, it changed nothing for Galileo personally. In the larger world, the Thirty Years' War had finished its first phase, the revolt of Bohemia against the Habsburgs having been crushed with the Habsburg victory at White Mountain (1620), but it was far from over. The elector Frederick V of the Palatinate, the Bohemian Protestant choice over the Catholic Habsburg Ferdinand II (1578–1637), had not abandoned his crown, and other powers were drawn into the vortex. Soon Spain and the Dutch Republic (who had a parallel war going on after the end of the Twelve Year Truce in 1621) would be drawn in, as well as, eventually, Denmark, France, and Sweden. The papacy, while hesitant in the beginning, soon threw its support behind the Habsburgs and their allies, the Catholic League, with substantial financial subsidies. Pope Gregory was much more aggressively Catholic than his predecessor, seeing the conflict in Germany as a holy war, and he increased the subsidies to the Habsburg emperor and the Catholic League, encouraging the Catholic princes in their re-Catholicization of the empire, even by force if necessary. After the fall of Heidelberg, the capital of Frederick V, the magnificent, world-famous Palatine

Library was given to the pope in gratitude for his aid, carried in fifty transport wagons across the Alps to Rome. The reign of Gregory XV was the apex of the Counter-Reformation and of Spanish influence in Rome, and it was appropriately symbolized in March 1622, when he canonized with the Italian Philip Neri four Spaniards (something for which the Spaniards had lobbied hard since the time of Philip II): the founder of the Jesuits Ignatius Loyola, the great Jesuit missionary Francis Xavier, the reformer of the Carmelites Teresa of Avila, and Isidore the Farmer, who had lived five hundred years earlier and had become a kind of patron saint of Madrid, the Spanish capital.

Pope Gregory XV also attempted to reform papal elections with his *Aeterni patris filius*, which regulated their procedures, and his *Decet Romanum pontificem*, which concerned the ceremonial aspects of the conclave. These documents made the rules of seclusion much stricter, guaranteed greater secrecy in the balloting, and made election by acclamation more difficult. While it made the jockeying and negotiations of the factions during the conclave somewhat more complicated and onerous, it did not end them. At the long, exhausting, and contentious conclave that followed Gregory's death in July 1623, the Florentine Maffeo Barberini was elected. While his election would produce a significant shift in Galileo's life, it also caused a momentous shift in political affairs, for Barberini was not a client and supporter of Spain as previous pontiffs had been but of France, Spain's archenemy.

While the election of Maffeo Barberini (1568–1644) in 1623 as Pope Urban VIII at the comparatively young age of fifty-five was something of a surprise, it was not so much of a surprise to him since when he was sixteen or seventeen and a student at Pisa, a certain Sicilian cleric, Don Andrea Lorestino, a dabbler in astrology later imprisoned by the Inquisition, had predicted that he would one day become pope. The deadlock between the Borghese and Ludovisi factions during the conclave offered him a unique opportunity, and despite the official accounts showing him piously uninvolved with the nitty gritty of a fiercely disputed papal election, Maffeo was not above the politicking and intrigue necessary to achieve power, at one point during the conclave even donning a false beard and wig to go to a secret meeting in the cell of Scipione Borghese, the head of the

Borghese faction. Urban saw himself as a man of destiny, a man chosen by God to be pope.[1]

The Barberini were originally from the small Tuscan town of Barberino di Val d'Elsa and had achieved some success as textile and
wool merchants in Florence for several generations. Maffeo, born
and raised in Florence, was the fifth son of a large family whose father
had died in 1571 when he was young. To start him on his ecclesiastical career he was sent at sixteen to live with his uncle, Monsignor
Francesco Barberini (1528–1600), an apostolic protonotary in the
Roman Curia who had secured lucrative offices for himself there and
had amassed a fortune. Maffeo finished his education at the Roman
College and studied law at the University of Pisa, where he lodged
with the future poet Michelangelo Buonarroti the Younger (1568–
1647), the great-nephew of the artist Michelangelo. As a protégé of
his uncle and the inheritor of his fortune, Maffeo worked his way
through the ecclesiastical system, becoming in 1604 an archbishop and
papal nuncio to France. Two years later he was made a cardinal at the
request of the French king Henry IV, and in 1617 he was made prefect of the *Segnatura di Giustizia*, an important post in the leading tribunal of the Roman Curia. He called his brother Carlo and his family
to Rome to share in his success and, as with many families, they
reinvented their past, changing the form of their name to something
more aristocratic-sounding and their coat of arms from three silver
horseflies on a red field to three golden bees on a blue field, dropping
the wool shears that had indicated the original source of their wealth.
They even sought to connect themselves to the fourteenth-century
poet Francesco da Barberino (1264–1348), to give the family some
intellectual and literary distinction.[2]

A man of refinement, culture, and intelligence, somewhat vain
about his appearance and his poetic ability, which was quite genuine,
he was also prone to fits of irrational anger, and if offended, his opposition could become implacable. His youth, energy, and intellectual

---

[1] Franco Mormando, *Bernini: His Life and His Rome* (Chicago: University of Chicago Press,
2011), 67–68.

[2] Janie Cole, "Cultural Clientelism and Brokerage Networks in Early Modern Florence
and Rome: New Correspondence between the Barberini and Michelangelo Buonarroti the
Younger", *Renaissance Quarterly* 60 (2007): 729–88.

curiosity promised a new age in the Eternal City and the Church at large. A friend of scholarship and the arts, he admired and desired the friendship of great men, whether of artists like Bernini, whom he treated as an intimate friend (smitten even) and showered with offices and commissions, or of scientists like Galileo, to whom he had been favorably disposed from early on. On October 2, 1611, while Maffeo was in Florence visiting his nieces who were Carmelite nuns there, he had attended a banquet at the court of Cosimo II, where Galileo debated floating bodies with a philosophy professor from Pisa, and Maffeo publicly took Galileo's side. This favor and friendship continued in the intervening years, with Maffeo sending him the Latin poem *Adulatio perniciosa* (*Dangerous Adulation*) in August 1620, brimming with fulsome praise, right up to a letter thanking Galileo for guiding his nephew Francesco Barberini to a successful completion of his doctoral studies at Pisa just weeks before Maffeo's election.

As with earlier papal families, Urban's family—his two surviving brothers and three nephews—became major beneficiaries of his success. Antonio (1569–1646), his youngest brother, was a strict and austere Capuchin who had walked to Rome for his brother's coronation. He became a cardinal in 1624 (wearing his Capuchin habit beneath his cardinal's robes), as did in 1623 his older nephew Francesco (1597–1679), who became at twenty-six Urban's main assistant. Later, in 1628, he made his much less suitable and morally doubtful youngest nephew, Antonio (1607–1671), a cardinal at nineteen, and from the beginning he showered his brother Carlo (1562–1630) and his middle nephew, Taddeo (1603–1647), with offices, pensions, and fiefs. Taddeo married into the ancient Roman noble family of the Colonna in 1627 and was himself ennobled, becoming Duke of Monterotundo and prince of Palestrina. Nepotism was expected, and it would have disturbed contemporaries if there had not been some, but the Barberini shocked even the hard-bitten Romans by their excess. On his deathbed even Urban had qualms of conscience over it. Francesco Cardinal Barberini amassed a fortune of sixty-three million scudi during his uncle's reign, while Taddeo amassed a somewhat lesser fortune of some forty-two million scudi. Taddeo was made lieutenant general of the church at twenty-three, and after the death of his father, Carlo, all of his father's offices fell to him: general of the Church, governor of the Borgo and Citiavecchia, castellano of

Sant' Angelo, captain of the papal guards. In 1631 Taddeo was made prefect of Rome, a hereditary office that would put his family ahead of all other papal houses, outraging Roman society. However, after his uncle's death and the election of a new pope, Taddeo and all the Barberini had to flee the Papal States for fear of arrest and imprisonment. While his family's assets would not be lost, and Taddeo's descendants would return to prominence in papal Rome, he himself would die in exile.

Pope Urban was not just a Francophile who owed his rise to the French crown (both the cardinalate and the papacy since he had the support of the French faction as well as the Borghese faction); he also wanted to free the papacy from a dependence on Spain and use France as a counterweight to enhance the papacy's power. One of the great events in the papal year—and one of the great manifestations of the power of the Spanish nation in Rome—was the reception of the Spanish tribute of seven thousand gold ducats and the *chinea* (a symbolic white Neapolitan horse) for the fiefdom of the Kingdom of Naples on the Feast of Saint Peter's. It was a grand ceremony with a large procession led by the Spanish ambassador and the members of the Spanish faction. Within the first year of Urban VIII's reign, when the event came up he instructed that the money and horse be given a day before the Vigil of Saint Peter's instead of on the feast and that the papal treasurer would receive the horse and money instead of the pope himself, as had been the custom. This deliberate snub was indicative of the shift in papal favor away from Spain, though Urban VIII continued the financial concessions to the Spanish Crown such as the *Cruzada*, *Excusado*, and *Subsidio*, the "Three Graces", the three taxes traditionally and periodically given to the Crown by the Holy See to defend the faith and fight the infidel. This was no small concession since the three graces and the other ecclesiastical levies were worth 3.68 million ducats a year by 1621, a third of the Crown's ordinary revenue.[3] After his election, believing that the Catholic powers had achieved their goals in Germany and the danger to the faith was over, he stopped the papal subsidies to the emperor and Catholic League—though he did restore some subsidies later. He also rebuilt

[3] Peter H. Wilson, *The Thirty Years War: Europe's Tragedy* (Cambridge, Mass.: Harvard University Press, 2009), 125.

the fortifications in Civitavecchia and other places, contrary to the 1557 treaty with Spain. The papacy was also much more indulgent to the French—even after their armies invaded Italy and occupied the Valtellina, which controlled one of the few passages from northern Italy across the Alps into Germany. Only the revolt of the Huguenots in 1625 forced the French to give it up and retire from Italy. In the 1630s especially, the Spanish and French factions angled, fought, and intrigued against each other seeking dominance in the papal city, at times even breaking out into deadly street battles.

In the second quarter of the seventeenth century, Rome came to the end of many years of growth and settled into a period of stability with a population of about 115,000 people. That being said, despite the vast new construction and renovations of these decades, despite the reorganization of the city and the population increases, the city still lived within its ancient Roman walls, where pigs still foraged and vineyards and gardens bloomed. The city was still dependent on religion and local agriculture for its flourishing, was still heavily male, was still socially highly stratified, with a large number of beggars, and was still lacking a strong industrial base with a majority of its population living on a subsistence level who could easily be thrown into destitution by the vagaries of life. Spiritually the city was in better shape than it had been before. In 1624 Pope Urban began a massive investigation or visitation that would last until 1632 and which produced a two-thousand-page report that revealed the vastly improved pastoral and spiritual health of his diocese. While the number of priests was about the same as other Italian cities, about 2 percent in the 1620s and 1630s (between 1500 and 2300), their moral lives and education had vastly improved. That being said, in the early 1630s when Count Giacinto Centini (1597–1635), the nephew of the cardinal Felice Centini, a cardinal inquisitor both during the 1615–1616 affair and the 1633 trial, was seeking assistance to clear his uncle's path to the papacy by killing Pope Urban by means of black magic, he was still able to find six priests who were not only willing to help him achieve this nefarious deed but who were also sufficiently conversant with the occult to be of service, several even being well-known necromancers. In 1635, before a crowd of twenty thousand, Centini was beheaded at the Campo de' Fiori (a privilege given him as a nobleman since his punishment was to be burned alive), while his two main priest conspirators, already defrocked, were hanged with him, the rest having been sentenced to the papal galleys.[4]

In 1626 Pope Urban would consecrate St. Peter's, though the expensive interior decoration was still to be done. Urban VIII's Rome was preeminently the city of the Baroque, and in his reign Gian Lorenzo Bernini was the master. Bernini's talents managed to transcend the vagaries of papal elections. Thus, he was patronized by the Borghese and by the Ludovisi who despised them, with Pope Gregory XV making him a knight in 1621 at twenty-three, which was why he was known as "Il Cavaliere" or "Cavaliere Bernini" from now on. Bernini's son Domenico wrote of his father after the election of Maffeo Cardinal Barberini: "On the very same day of his election, the new pope had the Cavaliere summoned to his presence and addressed him with these words: 'It is a great fortune for you, O Cavaliere, to see Cardinal Maffeo Barberini made pope, but our fortune is even greater to have Cavaliere Bernini alive during our pontificate.' "[5] Pope Urban settled on Bernini a slew of offices so that his work became part of the city, though the greatest was that, in 1629, of being the architect of St. Peter's, a position that Michelangelo had once held. In 1624, he was commissioned to do the Baldacchino in St. Peter's Basilica, a set of four sixty-foot-tall spiral bronze columns over the main altar and tomb of Saint Peter, which he completed in 1633 at a cost of two hundred thousand scudi.[6] The story went about that in order to get all the metal he needed, Urban had ordered the bronze ceiling of the Pantheon's portico stripped. This inspired the famous pasquinade about the Barberini, *Quod non fecerunt barbari, fecerunt Barberini.* "What the barbarians didn't do, the Barberini did." In fact, most of the bronze taken from the Pantheon was used to make cannon and very little was given to Bernini, who didn't use it anyway.[7] In St. Peter's, besides the Baldacchino and some lesser projects, he executed the adornment of the four great piers at the crossing of the nave, personally sculpting the *Saint Longinus* (one of the four

---

[4] Peter Rietbergen, *Power and Religion in Baroque Rome: Barberini Cultural Politics* (Leiden: Brill, 2006), 349–60.

[5] Mormando, *Bernini*, 70.

[6] Ibid., 84.

[7] Ibid., 85–86

giant statues, one for each pier), and the impressive tomb of Pope
Urban VIII (which he began in 1627 but only finished in 1647). The
great Colonnade and Piazza of St. Peter's and the Chair of Peter
would be done under Pope Alexander VII (1655–1667). Bernini was
held in such great esteem by Pope Urban VIII that when he fell seriously ill in 1635, and came close to death, he was not only visited by
the three Barberini cardinals but even by Pope Urban VIII himself in
a great retinue and with much pomp, an unheard-of thing in Rome.
When Urban VIII died in 1644, he was despised and the Barberini
hated, so much so that they had to flee the Papal States; but Rome
without Bernini is unthinkable, and for that we can largely thank
Maffeo Barberini.

After Urban VIII's election Galileo was urged to come to Rome.
Hearing from Ciampoli, Cesi, and other high-placed friends that his
visit to Rome and to the pope would be both appreciated and propitious, he decided to go. Ciampoli was particularly well-placed
within the pope's entourage, and he passed on to Galileo that he read
to the pope selected passages of *The Assayer* at his meals and that the
pope had enjoyed them immensely. After a brief illness and some bad
weather Galileo did visit, though in the capacity of a private citizen,
even if he was armed with letters of recommendation from the Grand
Duke and the Grand Duchess Christina.

In 1624, from April 23 to June 16, Galileo was in Rome, where he
was well received by the pope and showed every sign of respect, having six audiences with the pope, who left him with parting gifts and
favors. While we are not certain of the exact content of their conver-
sations, Galileo returned home encouraged that he could write about
Copernicanism as long as he did so as something hypothetical and
not as something true. The difficulty was that "hypothetical" meant
different things to different people. For Galileo it was an assumption
that would not only be useful but, in fact, might be true even if it
were not as yet proven. For Pope Urban VIII, however, it was something purely instrumental, used as a means of prediction and calcula-
tion, a useful fiction rather than a real attempt to describe the world.

Another aspect to Urban VIII's worldview, and he would demand
it be included in Galileo's great work on cosmology, and one to
which we have already briefly referred, was his profound belief that it
was not possible for man to plumb the mysteries of the world and that

any attempts to do so were attempts to limit God's person and power. Human reason, he believed, was limited, and God being God could always create a world in a way different from our expectations. This extreme voluntarism (sometimes called nominalism) which exalted God's will over his reason had its roots, as we saw earlier, in the reaction against the naturalism of Aristotle in the late Middle Ages and was an attempt to protect God's freedom and transcendence. It was also a long-standing position of Urban VIII and one well suited to an age much influenced by skepticism, as the great popularity of Montaigne's *Essays* also indicates. In *De Deo Uno* (1629), a book by his personal theologian ("my Bellarmine") and member of his intimate circle, Agostino Oreggi (1577–1635), and dedicated to the pope, on the question of whether God knows future contingents (we are back once again to the *De auxiliis* controversy), Oreggi recounts a conversation, probably from 1615 or 1616, between a then Cardinal Barberini and a scientist who could only be Galileo.

I began to pay more carefully attention to the value to be given to the proceeding argument, while the Supreme Pontiff Urban VIII (whom God may preserve safe and sound to his Church for a long time), when still a Cardinal, was admonishing his friend—no less illustrious for his science than praiseworthy for his religious faith—to turn his mind carefully to the question whether those things agree with the Holy Scriptures which he had thought out concerning the earth's motion in order to save all the phenomena appearing in the sky, as well as anything that the philosophers, as a result of their careful inspection and observation, generally admit about the movements of the sky and stars. Having conceded everything that the most learned man had thought out, he asked if God would have had the power and the knowledge to arrange and move differently the orbs and the stars in such a way as to save all the phenomena that appear in the skies, as well as everything reported on the motions, order, location, distance, and arrangement of the stars.

If you deny this, said Sanctissimus [Barberini], then you must prove that for things to happen otherwise than you have thought out implies a contradiction. For God in his infinite power can do anything which does not imply a contradiction; and since God's knowledge is not inferior to his power, if we admit that God could have done so, then we have to affirm that he would have known how. And if God had the power and the knowledge to arrange these things differently from

the way thought out, while saving all that has been said, then we must
not restrict to this way the divine power and knowledge.

Having heard these words, the most learned man remained silent,
thus deserving praise for his virtue no less than for his intellect.[8]

Rivka Feldhay has remarked on what she sees as the similarities
of the skeptical arguments of Oreggi (and of Pope Urban) to those of
the great Dominican of the *De auxiliis* controversy, Thomas de Lemos,
and posits that this new skeptical sensibility (so unlike the traditional
Thomistic view) was a product of that fierce controversy on grace
and freedom which was the *De auxiliis* conflict.[9] This may be so,
especially considering the question in which it is brought up, but we
will have to leave that to others to investigate.

In any case, Galileo must have been aware of what Urban meant
by hypothetical and must have realized the extreme importance to
Urban of this argument. His seemingly cavalier treatment of these
concerns would cause great trouble later on, leading to a profound
sense of betrayal in Urban when *The Dialogue Concerning the Two
Chief World Systems* was published in 1632. But he presumed he could
finesse it, and by faithfully keeping to the letter of what was demanded
of him he could achieve his larger aims. While Galileo well knew he
could not demonstrate the truth of Copernicanism (which is why
he was forced to go through all these hoops to discuss it at all), it is
not very difficult to see that he went out of his way to make it far
more plausible than anything else. He should not have been surprised
at the pope's negative reaction, though the intensity and duration of
it, I think, might have surprised him. Galileo overestimated his charm
and his skill, and he would pay a terrible price.

The idea for the *Dialogue* goes back many decades, and it was
mentioned both in the *Starry Messenger* and in the list of works that
Galileo proposed to the Grand Duke when he sought to return to
Florence and serve as his official "mathematician and philosopher".
But before he began working on the *Dialogue* he spent a few months
writing his *Reply to Ingoli* as a trial balloon to see how far he could go

---

[8] Jules Speller, *Galileo's Inquisition Trial Revisited* (Frankfurt: Peter Lang, 2008), 376–77.

[9] Rivka Feldhay, *Galileo and the Church* (Cambridge: Cambridge University Press, 1995),
208–11. See Speller, *Galileo's Inquisition Trial Revisited*, 375–83.

in his promotion of Copernicanism. Francesco Ingoli (1578–1649), now secretary to the newly created Congregation for the Propagation of the Faith, had debated the topic of Copernicanism with Galileo in 1616, writing a letter addressed to him showing his reasons for opposition. But Galileo could not respond because of the decision that year against heliocentrism. His belated, fifty-page response (which like the *Dialogue* contrasted the Ptolemaic and Copernican systems) was sent around to friends who generally praised it, but by April 1625 it was decided not to send it to Ingoli or publish it since Galileo's *Assayer* had been denounced some months earlier to the Inquisition, and Cardinal Barberini, a member of the Linceans since 1623 and their great protector, had been sent to France on a diplomatic mission and could not be around to defend it. While Galileo's friends thought the denunciation was for Copernicanism, in fact the denunciation accused him of being an atomist and so, ultimately, denying the Council of Trent's pronouncements on the Holy Eucharist.

Atomism, which viewed matter as made up of invisible tiny particles or atoms, went back to the ancient Greek philosopher Democritus (c. 460–c. 370 B.C.), and with the revival of Epicureanism in the sixteenth and seventeenth century, it returned; and in the form of corpuscularism (a type of atomism but with the particles being divisible not indivisible), it would form the foundation of the mechanical philosophy so influential in the Scientific Revolution. While many saw it as highly dangerous to the Catholic belief in the Holy Eucharist because of its denial of the ideas of substance and accidents, the philosophical terms used to describe the transformation ("transubstantiation") of the bread and wine into Christ's Body and Blood, there were also sincere Catholic scholars such as Pierre Gassendi (1592–1655) who were heavily influenced by it. While there is some element of corpuscularism in Galileo, it was a small part of the work; and when Francesco Cardinal Barberini, a member of the Inquisition, delegated the investigation to his personal theologian, Giovanni di Guevara, he saw nothing in it and eventually dropped the matter.

With this reference to the atomism of *The Assayer*—which would also be discussed by Grassi in his rejoinder to Galileo of 1626, the *Ratio ponderum*, and again come up during Galileo's 1633 trial—we have an appropriate place to discuss one of the most radical inversions of the Galileo case, that of Pietro Redondi's *Galileo Heretic* (1983), which caused such a stir some years ago. It was Redondi's thesis that the trial of Galileo was orchestrated by Pope Urban, not to punish Galileo for his Copernicanism but to protect him from the charge of atomism by organizing a show trial using the lesser charge of his Copernicanism as the cover. If this were true, it would indeed be a most extraordinary turn, but historians have pointed to the profound lack of evidence for any of this and the overwhelming evidence that Copernicanism was the point of conflict. While the Holy Eucharist was certainly a pillar of the Catholic faith, and while there was certainly some concern over the possible implications of atomism to undermine the Catholic doctrine of Transubstantiation, it did not rank high among the concerns of the actors of this drama. Historians have pointed out Redondi's systematic distortion of facts, his highly selective use of facts, and the remarkably weak arguments and fanciful conjectures that make up the narrative; and, however admirable the book may be in some respects, it has been seen as more historical fiction than history.[10]

By the end of 1625 Galileo was writing that the *Dialogue* was almost finished, but it then made slow progress for a number of reasons: because he was often sick and could not devote much of his diminished energies to writing; because family issues took up time, with the return of his brother Michelangelo from Munich with his family, and with his own son Vincenzo's difficulties; and because his attention and focus were not what they used to be. Galileo was also always aware that there was strong opposition in Rome to him and to what he wanted to achieve—which could only further his hesitancy. In March 1629 he became deathly ill but recovered. In October 1629 he was able to return to work after a three-year hiatus, and by January 1630 he had finished the book. The fact that Riccardi, who had so extravagantly praised *The Assayer*, had become in 1629 the Master of the Sacred Palace, the man in charge of licensing the book, could only have been a further impetus since he was assured of a favorable response. Originally wishing to mention in the title

[10] Owen Gingerich, "Show Trial?" *The American Scholar* 59, no. 2 (Spring 1990): 310–14; Vincenzo Ferrone and Massimo Firpo, "From Inquisitors to Microhistorians: A Critique of Pietro Redondi *Galileo eretico*", *Journal of Modern History* 58 (1986): 485–524; Richard S. Westfall, *Essays on the Trial of Galileo* (Notre Dame, Ind.: University of Notre Dame Press, 1989), 84–103.

the tides (which he considered the strongest physical evidence for Copernicanism), he dropped that idea when Pope Urban objected, though he kept the argument in where it absorbs the entire discussion of the Fourth Day.

The *Dialogue of the Two Chief World Systems* was a four-day discussion set in Venice between Filippo Salviati, Giovanfrancesco Sagredo, and an Aristotelian professor called Simplicio. The first two were old friends of Galileo now long dead: Salviati, a fellow Florentine and Lincean, dying in 1614, and Sagredo, his boon companion from his Padua days, dying in 1620. Simplicio was a fictional character most likely named after the Greek philosopher and commentator on Aristotle of the sixth century A.D., Simplicius of Cilicia. That the name was also similar to the Italian word for simpleton was surely an added advantage. Certainly, he comes across in the dialogue as neither very bright nor very persuasive, a buffoon who supplies the standard philosophical objections to Copernicanism. In him we see bits of Galileo's friend, the strict Aristotelian Cremonini, as well as the leader of the Pigeon League, Ludovico delle Colombe, Galileo's fiercest philosophical enemy. Salviati is clearly the defender of Copernicanism, while Sagredo, in whose palace the discussions occur, appears as a neutral, open-minded, and educated person, though somewhat inclined to Copernicanism. The text is not an arid work of science but a literary masterpiece infused with all of Galileo's brilliant artistic sensibility, with his apt analogies and insightful descriptions, and with his sharp satire and irony.

Galileo visited Rome from May 3, 1630, to June 26, 1630, a guest of the Tuscan ambassador Francesco Niccolini at the Palazzo Firenze. He met with Pope Urban VIII on May 18. He also met with Niccolò Riccardi, the Master of the Sacred Palace and the person in charge of reading and approving his book for publication. Riccardi (1585–1639), though born in Genoa, had joined the Dominicans in Spain as a young man. He was educated there and gained a reputation as a great preacher, which earned for him from King Philip III of Spain the name of *Padre Mostro* (Father Monster), descriptive both of his vast girth and his vast erudition and eloquence. Father Monster, as everyone called him, was also a cousin of Caterina Niccolini, the wife of the Tuscan ambassador. Riccardi had Raffaello Visconti, a professor of mathematics and a Dominican colleague, read and review the manuscript. He made some minor changes and approved it. Riccardi then also approved it, asking only for minor corrections before Galileo brought it back to Rome for final approval. Riccardi spoke with the pope, who gave his approval, and before Galileo left he had a genial meeting with Pope Urban and dinner with Francesco Cardinal Barberini. Galileo's intention after he returned home was to finish the manuscript, return, get final approval with the help of Ciampoli to resolve any differences, and then give it to Prince Cesi, to be printed under the auspices of the Linceans. Then everything seemed to go wrong.

The first of these difficulties was the death of Prince Cesi on August 1, 1630. Besides losing a valuable friend who was well connected in Rome and understood how the system worked, whom he had often asked for advice on what was possible and how he should proceed in the labyrinth of the Eternal City, he also lost the man who could have brought the manuscript to a successful printing with all the possible glitches smoothed out. Without his support Galileo would have to make other arrangements for the book's publication.

The second of these complications was the outbreak of the plague that struck northern Italy in 1629. In 1627 Duke Vincenzo Gonzaga of Mantua and Monferrato died without issue, and the War of the Mantuan Succession (1628–1631) broke out, with France supporting one candidate and the Spanish and the Austrians another. Rival armies devastated parts of northern Italy, with the Imperial forces from the north not only barbarically sacking the great city of Mantua, but also bringing plague with them. Described by Manzoni (1785–1873) in his great novel *The Betrothed*, it would kill a third of the population of northern Italy. It reached Tuscany in the summer of 1630, killing seven thousand in Florence alone, and any traffic between Rome (where the plague did not strike) was heavily impeded because of the severe quarantine. Even a manuscript might be at risk of being confiscated and destroyed. The spread of this plague would make it difficult for Galileo to send his revisions to Rome for final approval, and would create the confusing situation where part of the *Dialogue* was approved by Riccardi in Rome and part by another in Florence. That would play an important part in Galileo's later difficulties.

The third difficulty touched upon the political situation. Pope Urban had tried to keep a balance between the two great Catholic

powers of France and the Habsburgs, but the situation became unten-
able. In 1629 the Catholic cause in the Thirty Years' War finally
seemed at the point of total victory. Emperor Ferdinand II had not
only defeated Frederick V of the Palatinate and the Protestant Union,
reconquering Bohemia and dismembering Frederick's lands, but he
had defeated the intervention of the Protestant king Christian IV
of Denmark, and, under Albrecht von Wallenstein (1583–1634), the
Imperial forces had subdued most of northern Germany. In March
1629 Ferdinand promulgated the Edict of Restitution, which would
have restored to the Church possessions seized by the Protestants
since 1552, contrary to the conditions, the emperor said, of the
Treaty of Augsburg of 1555, which had ended the religious war in
Germany: two archbishoprics and thirteen bishoprics in northern
Germany and more than five hundred monasteries, mainly in Lower
Saxony, Württemberg, and Franconia, would have to be returned.[11]
In July 1630, encouraged by the French who would later help finance
him, Gustavus Adolphus (1594–1632), king of Sweden, invaded Germany. On September 17, 1631, at Breitenfeld just four miles north
of Leipzig, the Catholic League army was annihilated, soon to be
followed by other Swedish victories and conquests. By the spring of
1632 Swedish troops were ravaging Catholic Bavaria, and an invasion
of Italy seemed a real possibility.

After Breitenfeld, the pressure on Urban to support the Catholic forces more extensively grew ever stronger, including specific
demands that he join the Catholic League, send papal troops to
Germany, allow for new and substantial subsidies from the Church
in Spanish lands to support the Crown's military efforts, and even
use the spiritual weapons of excommunication and interdict against
those that supported the Protestants such as France. While Urban
did allow for some short-term support from some Church lands,
he ignored the rest, inflaming the already-strong Spanish opposition to him. Already in 1630 street fights were breaking out in
Rome between the supporters of Spain and France, and Spanish
anger came publicly to the surface in a consistory of cardinals on
March 8, 1632, when the Spanish ambassador Cardinal Gaspar de
Borja y Velasco (1580–1645), grandson of the third Jesuit general

[11] Wilson, Thirty Years War, 448.

Saint Francesco Borja and sixth son of the Duke of Gandia, publicly attacked the pope for failing to support the Spanish king in his war against the German Protestants and blaming the pope for whatever losses the Catholic faith might suffer. Swiss Guards had to be brought in to restore order among the cardinals before real violence could erupt. Cardinal Borja gave the widest publicity to the protest in the city, and there was talk on the Spanish side of invasion, of a general council, and even of a papal deposition, threats that Urban had to take very seriously.

While it is important to remember the heightened sense of crisis and the tension this would cause in Rome and for Urban personally, and even the personal need to reassert his authority in face of such a crisis, I do not think, as some do, that Pope Urban's about-face in reference to Galileo, his trial and condemnation, can be seen as a political play by Urban to appease the Spanish, to undercut their charges and to prove his Catholic bona fides, as his own Catholicism was being challenged. As Westfall points out, we see nothing in the documentation and diplomatic correspondence that the Spanish considered Galileo as a grievance.[12] Their grievances all center around Urban VIII's perceived support of France and insufficient support of Spain. In fact, Galileo had been patronized by both the Austrian and Spanish Habsburgs, the main pillars of the Church against the Protestants, with King Philip IV of Spain in 1629, just as Galileo was completing the *Dialogue*, even writing to Galileo through the Florentine court asking for one his special telescopes. Galileo's main patrons, the Medici Grand Dukes of Tuscany, were also known as defenders of the Church and supporters of the Habsburgs. (The 1608 marriage of Cosimo II to the Habsburg Maria Maddalena of Austria had realigned the Medici back to the Habsburgs.) Attacking a favorite of the Catholic powers would seem to me an odd way of showing one's orthodoxy. And as Westfall has pointed out, Pope Urban continued his seemingly anti-Spanish policies until 1635, long after the Galileo case was finished.

While there was little that Urban could do against Borja and the Spanish cardinals, backed as they were by the Spanish king, there was much he could do against their Italian supporters. Cardinal Ubaldini, who had had previous run-ins with Urban, was threatened with

[12] Westfall, *Essays on the Trial of Galileo*, 92–95.

imprisonment in Castel Sant' Angelo and was forced to write an apology. Cardinal Ludovisi, the once all-powerful cardinal-nephew of the previous pope, Gregory XV, was told to return to his archdiocese of Bologna within ten or twelve days, and if he did not go willingly, then he would be obliged to do so by force. He died there nine months later. Another casualty of Borja intervention was Galileo's friend Giovanni Ciampoli who was exiled in November 1632 to provincial governorships never to return.[13]

The final blow concerned astrology. While Galileo was in Rome, the city was full of rumor about a recent horoscope foretelling the death of the pope and of his nephew, the family heir, Taddeo Barberini. So strong were the rumors of Urban's imminent death that King Philip III sent the Spanish cardinals to Rome to prepare for a conclave, soon followed by those of the French and Germans. More dangerously, Galileo began to be associated with the horoscope. A Roman *avviso* (a handwritten newsletter that was the forerunner of the newspaper) of May 18, 1630, announced:

> Galileo, the famous mathematician and astrologer, is here [in Rome] to try and publish a book in which he attacks many opinions held by the Jesuits. He has been understood to say that D. Anna [that is, Anna Colonna, wife of Taddeo Barberini, the pope's nephew] will give birth to a son, that we will see peace in Italy at the end of June, and that shortly thereafter Taddeo and the pope will die. This last point is confirmed by the Neapolitan Caracioli, by Father Campanella and by several writings that discuss the election of the new Pontiff as if the Holy See were already vacant.[14]

Galileo was so concerned about this that he had his friend Michelangelo Buonarroti the Younger speak to Francisco Cardinal Barberini to plead Galileo's innocence. Knowing as he did that the pope had a profound belief in astrology, Galileo was relieved to hear that the rumors connecting him with the horoscope were not believed by the pope. Urban VIII was quite adept at astrology and as pope even had horoscopes done of the cardinals and would openly predict

---

[13] Ibid., 96.

[14] William Shea and Mariano Artigas, *Galileo in Rome: The Rise and Fall of a Troublesome Genius* (Oxford: Oxford University Press, 2003), 139.

the times of their deaths. The Venetian ambassador from 1632 to 1635, Alvise Contarini, wrote of Pope Urban: "He lives by strict rule; he regulates his actions in great part according to the movements of the heavens, about which he is very knowledgeable, even though he has with very great censure prohibited its study to everyone else."[15] Beginning in 1626 rumors and predictions of his own imminent death had first begun to spread, and they became overwhelming by 1628, actively encouraged by the Spanish. There were two particularly dangerous moments: in 1628 there was an eclipse of the moon in January and of the sun in December, and in 1630 there was a solar eclipse in June. Terrified, Urban VIII called upon the services of Tommaso Campanella, who had been sent from the prisons of Spanish-controlled Naples in 1626 to the prison of the Inquisition in Rome and was known as an expert on astrology. In July 1628 he was released from prison to detention in the monastery of Santa Maria sopra Minerva, the headquarters of the Dominican order, and during the summer of 1628 the Tuscan and Venetian ambassadors reported frequent secret meetings between Campanella and Pope Urban. On January 11, 1629, Campanella was completely free. Heavily indebted to Ficinian magic, his *De siderali fato vitando* (On Avoiding the Fate of the Stars) contained a description of what countermeasures one could utilize against these negative astral forces, particularly the evil influences of Mars and Saturn, and it is very likely that the ceremony described there is what was performed for Pope Urban.

Dressed in white robes, having locked themselves in a room with the doors and windows sealed against the infections of the poisoned air, they would first sprinkle the room with rose vinegar and other aromatic substances, and then burn laurel, myrtle, rosemary, cypress, and other aromatic herbs. Then the room was to be hung with white silken cloths and branches, and decorated with the twelve signs of the zodiac. Two candles and five torches would be lit, symbolizing the seven planets, to provide a correct substitute to heavens made defective by the eclipse. All the while jovial and venereal music (music of a happy, sweet character associated with the beneficent planets of Jupiter and Venus) was to be played to

[15]John Belton Scott, *Images of Nepotism: The Painted Ceilings of Palazzo Barberini* (Princeton, N.J.: Princeton University Press, 1991), 86.

disperse the evil qualities of the eclipse-infected air and to expel the influence of the bad planets. They would also utilize stones, plants, colors, and odors belonging to the good planets of Jupiter and Venus to do the same and imbibe astrologically distilled liquors.[16]

In February 1630, Urban's brother Carlo died and Pope Urban was said to be relieved. It seems that Campanella's services were again used in December 1630 when the son of Taddeo Barberini was suffering from another such bad astral influence. For a while Campanella's star shined brightly. It seems he even attempted to be made a consultor of the Inquisition, but he lost Pope Urban's favor over the publication in 1629 of *De fato siderali vitando*, which Campanella had not intended to publish. He blamed this on the malice of two highly placed Dominicans (they were unnamed but they seem to have been Niccolò Ridolfi, Master of the Sacred Palace and later Master General of the Dominican order, and Niccolò Riccardi) who had arranged the book to be published, giving the impression that it was with Campanella's permission, in the hope of embarrassing both Urban and Campanella. While he was no longer as intimate with Pope Urban, when in 1634 the Spanish government wanted to extradite Campanella as a continuing danger to their government in Naples, Urban still connived with the French ambassador to spirit him out of Rome, disguised as a Minim friar, under a false name and carried in the coach of the French ambassador. A monthly pension was paid to him even after he fled to France and the Bourbon court.

To this already-heightened atmosphere of crisis and danger caused by the political and astrological situations we have described, something truly shocking was added. In July 1630, the head of the Roman monastery of Santa Prassede and onetime general of the Vallombrosian order, Abbot Orazio Morandi, was arrested and imprisoned for producing the recent horoscope which had caused so much trouble. Under Morandi the monastery had become a major center in Rome for information of every sort. It possessed a formidable library (including a very extensive collection of forbidden books and manuscripts, including the forbidden *avvisi* which carried all the latest news

---

[16] Ibid., 74; and D. P. Walker, *Spiritual and Demonic Magic from Ficino to Campanella* (Notre Dame, Ind.: Notre Dame University Press, 1975), 207.

and gossip) and, in a time when lending libraries did not really exist, it was generous in allowing select persons to take out books. Its clientele included many influential and highly placed figures, of Roman aristocrats, of cardinals and leading prelates in the Roman Curia, of diplomats such as the Venetian and Tuscan ambassadors, and of artists, writers, and other important cultural figures. The monastery was also a major center of astrology with Morandi carrying the reputation as being the most sought after astrologer in Rome. His goal was to reform astrology by basing it on experience and observation, and he was working on a vast encyclopedia of astrology that would have put it on a scientific footing. When the magistrates searched the monastery, they found a veritable "astrological database", with dozens of nativities (natal horoscopes) including those of Pope Urban and many cardinals. Morandi it seems had even gone so far as to prepare for the Venetian ambassador an astrological analysis of the various cardinals who might succeed Urban, giving their strengths and weaknesses, which he embellished with portraits. Among these discovered horoscopes was also that of Galileo, a friend and correspondent of Morandi, who had dined there just a few weeks earlier when he brought the manuscript for his *Dialogue on the Two Chief World Systems* to Rome for approval. The Morandi circle was extensive, and conveniently for everybody involved, Morandi would unexpectedly die in prison in November, of natural causes it seems, bringing an abrupt conclusion to a trial that could only have embarrassed not a few highly placed figures in Rome. While most in the Morandi circle escaped unscathed, Raffaello Visconti, who had given approval to publish Galileo's *Dialogue*, did not. Though he admitted that he had discussed Urban's nativity with Morandi, he defended himself by telling the investigators that his interpretation of the nativity and his response to Morandi was that Urban would live a long life, with 1643 or 1644 being the dangerous years for him (Urban died in 1644). His defense did not save him, however, and he was exiled in disgrace to Viterbo in the Papal States.

Galileo's association with Morandi and Visconti could not have helped his cause. Nor was Galileo unfamiliar with the Roman political astrological culture. Galileo, it seems, was instrumental in translating and publishing in 1626 a prognostication that Portugal (which had been conquered by Spain in 1580) would soon be free of Spanish control, to which he added a prefatory letter praising its astrological contents.[17] While it was not the most accurate of predictions since Portugal would not regain her independence until 1640, it certainly reveals an unknown aspect to Galileo's career and his connection to Rome's highly dangerous and politicized astrological world, though one in this case on the anti-Spanish side.

In April 1631 Pope Urban published *Inscrutabilis*, his famous bull against astrology, using his standard argument of man's incapacity to fathom the awesome and impenetrable mystery of God. That this bull was a response to his own situation is clear from the fact that the only practices it specifically condemned were the predictions of the deaths of princes and the pope, including their family members up to the third degree of consanguinity, which it condemned under penalty of death and confiscation of goods.

The unfathomable depth of God's judgements does not tolerate the human intellect, locked as it is into the shadowy prison of the body, to rise above the stars. Not only does it dare in its sinful curiosity to pry the mysteries which are hidden in God's heart and are unknown even to the spirits of the blessed, it also presumes, arrogantly, to set a dangerous example and hawks mysteries as certainties, scorning God, agitating the state and endangering princes.[18]

Galileo finally received permission in September from Riccardi to print his *Dialogue* in Florence, but Riccardi still wanted to see the whole manuscript first. Fearing the manuscript might get damaged or lost in transit, through the intervention of the Tuscan ambassador's wife, Riccardi's cousin, Galileo received permission in October to send just the preface and conclusion, but the book had to be revised by a seasoned Dominican theologian in Florence. Galileo suggested Giacinto Stefani, a consultant to the Florentine Inquisition and former court preacher to the Grand Duchess Christina. Riccardi agreed in November and gave Stefani some instructions on what was

---

[17] H. Darrel Rutkin, "Galileo Astrologer: Astrology and Mathematical Practice in the Late-Sixteenth and Early-Seventeenth Century", *Galilaeana* 2 (2005): 136.

[18] Germana Ernst, "Astrology, Religion, and Politics in Counter-Reformation Rome", in S. Pumfrey, P. L. Rossi, and M. Slawinski, eds., *Science, Culture, and Popular Belief in Renaissance Europe* (Manchester: Manchester University Press, 1991), 271.

needed. Galileo sent the preface and conclusion but heard nothing for months. On March 7, 1631, he wrote to the Tuscan Secretary of State since 1627, Andrea Cioli (1573–1641), about the delay, telling him that months earlier he had heard from Castelli that Riccardi had told him a number of times that he was going to return the preface and conclusion, revised to his complete satisfaction, but he never did. The Grand Duke put pressure on his ambassador in Rome, Niccolini, who put pressure on Riccardi, who was clearly dragging his feet. In a letter of April 18, Niccolini admitted to Cioli that Galileo's opinions were not welcome in Rome, especially by the authorities. On April 25 Riccardi replied to Niccolini, explaining that he only wanted to prevent harm to Galileo's reputation and that he could not give the imprimatur to a book printed outside his jurisdiction, though he could ensure that the revisions followed the conditions set by the pope by writing to the Florentine inquisitor about them. Only on May 24 did he finally send the papal conditions, saying it was up to Stefani to approve the book or not; and only on July 19 did he send the preface (which he could change as long as he kept the substance), and no suggested text for the conclusion except for the instruction that it should match the preface and include the argument from divine omnipotence that the pope gave him. There were three conditions: first, that the title and subject of the work not focus on the ebb and flow of the tides but absolutely on the mathematical examination of the Copernican position, with the aim of proving that absent divine revelation the appearances can be saved by this position; second, that the absolute truth of this opinion would never be admitted but only its hypothetical truth; third, that it must be shown that it was not by a lack of knowledge on the Church's part that Copernicanism was condemned in 1616, but that they were aware of all the arguments in its favor. Riccardi assured them that if these provisions were kept, the book would encounter no obstacle in Rome.

On February 21, 1632, after nine months, the Florentine printer Giovani Battista Landini finished printing a thousand copies of the book, a large run for the time. The preface was printed in italics and in a different typeface from the rest of the book—something that later critics would object to as making it seem something alien and added on to the text.

Several years ago there was published in Rome a salutary edict which, in order to obviate the dangerous tendencies of our present age, imposed a reasonable silence upon the Pythagorean opinion that the earth moves. There were those who impudently asserted that the decree had its origins not in judicious inquiry, but in passion none too well informed. Complaints were to be heard that advisors who were totally unskilled at astronomical observation ought not to clip the wings of reflective intellects by means of rash prohibitions.

Upon hearing such carping insolence, my zeal could not be contained. Being thoroughly informed about that prudent determination, I decided to appear openly in the theater of the world as a witness of the sober truth. I was at that time in Rome; I was not only received by the most eminent prelates of that Court, but had their applause; indeed, this decree was not published without some previous notice of it having being given to me. Therefore I propose in the present work to show to foreign nations that as much is understood of this matter in Italy, and particularly in Rome, as transalpine diligence can ever have imagined. Collecting all the reflections that properly concern the Copernican system, I shall make it known that everything was brought before the attention of the Roman censorship, and that there proceed from this clime not only dogmas for the welfare of the soul, but ingenious discoveries for the delight of the mind as well.[19]

Galileo went on to say that he would take on the Copernican way of thinking as a purely mathematical hypothesis, striving to show its superiority to the view of the motionless earth relative to the arguments of the Peripatetics, or at least of those who claim that name, however poorly they philosophize. He would be treating three points. First, that he would try to show that any experiments possible here on earth were insufficient to prove its mobility and could be adapted to either a moving or a stationary earth. Second, that he would examine the phenomena of the stars strengthening the Copernican hypothesis so that it would appear to emerge absolutely conclusive, adding new speculations, though these were advanced for the sake of astronomical usefulness and not to imply them as certainly real. Third, to see if the problem of the tides could tell us something about the earth's

---

[19] Galileo Galilei, *Dialogue on the Two Chief World Systems*, trans. Stillman Drake (New York: Modern Library, 2001), 5.

motion. So that no foreigner could ever accuse him of ignoring this important reality of the tides, he would make the argument from the tides for the motion of the earth seem as plausible. He hoped to show by these discussions that Italians were no less thoughtful than other nations and that the assertion of the stationary earth was not based on ignorance but on piety, religion, an acceptance of an all-powerful God, and an awareness of human limitations.

Besides the preface, which is exactly as Riccardi sent except for one word, there were frequent disclaimers throughout the text of the hypothetical nature of Copernicanism, and at the end, after a discussion of the argument of the tides, was placed Pope Urban's doctrine of divine omnipotence, "the medicine at the end", though placed in the mouth of Simplicio. Having just heard Salviati's argument supporting a moving earth from the tides, Simplicio remarks that he really did not completely understand it and does not believe it.

> I do not therefore consider them true and conclusive; indeed, keeping always before my mind's eye a most solid doctrine that I once heard from a most eminent and learned person, and before which one must fall silent, I know that if asked, whether God in His infinite power and wisdom could have conferred upon the watery element its observed reciprocating motion using some other means than moving its containing vessels, both of you would reply that He could have, and that He would have known how to do this in many ways which are unthinkable to our minds. From this I forthwith conclude that, this being so, it would be excessive boldness for anyone to limit and restrict the Divine power and wisdom to some particular fancy of his own.[20]

Salviati then replies before they finally end their discussions and break up to go on a gondola ride:

> An admirable and angelic doctrine, and well in accord with one another one, also Divine, which, while it grants to us the right to argue about the constitution of the universe (perhaps in order that the working of the human mind may not be curtailed or made lazy) adds

[20] Ibid., 538.

that we cannot discover the work of His hands. Let us, then, exercise these activities permitted to us and ordained by God, that we may recognize and thereby so much the more admire His greatness, however much less fit we may find ourselves to penetrate the profound depths of His infinite wisdom.[21]

While Galileo gave the pope's argument its full due, he undermined the pope's argument, however, not only by placing it in the mouth of the character who had the least credibility of the three, and was even in some sense ridiculous (to whom else could he have given it realistically?), but also by placing throughout the text statements from Salviati that seemed to undermine the argument, which in effect said that God always acted in the simplest way possible and not through miracles when natural causes would be sufficient.[22] Thus, for example, near the beginning of Book Four when Simplicio introduces the idea that the movement of the tides might be supernatural and miraculous, by the absolute power of God, Salviati turns the argument on its head by saying that if we must introduce a miracle to produce the ebb and flow of the tides, then why not just have the earth move miraculously so as to produce the movement of the seas, which is far simpler and more natural, covers better the sea's variety (which involves many different miracles), and removes the necessity of using a miracle of keeping the earth stationary against the sea's powerful impulses.

While Salviati admitted that the wisest man of Greece (Socrates) openly said that he was aware that he knew nothing, Salviati also said that Socrates still knew he had limited wisdom—which is something even if it is nothing compared to infinite wisdom. Salviati said that a sharp philosophical distinction should be made, with understanding as being taken in two ways, intensively or extensively. Extensively, in regard to the infinite number of intelligible things, we know nothing since even a thousand propositions compared to infinity is like nothing. Intensively, insofar as we know an individual proposition, we know some propositions perfectly and with absolute certainty. This is true for the purely mathematical sciences of geometry

---

[21] Ibid., 538–39.
[22] Winifred Lovell Wisan, "Galileo and God's Creation", *Isis* 77 (1986): 481.

and arithmetic. God knows more propositions since he knows them all, "but for the few understood by the human intellect, I believe our knowledge equals the divine one in regard to objective certainty, for it is capable of grasping their necessity, which seems to be the greatest possible assurance there is."[23]

Simplicio thought this a "very serious and bold manner of speaking", and the special commission set up by Pope Urban to investigate the work after its publication thought so too, for they included this criticism in their report: "That he wrongly asserts and declares a certain equality between the human and divine intellect in the understanding of geometrical matters."[24] Among the many things over which Pope Urban could have felt betrayed, it seems that it was Galileo's attack on the pope's deep-seated skepticism that was the most important. His putting the human mind in equality with God and forcing necessity onto God struck at the very core of Urban's worldview. To my mind, it best explains the ferocious and implacable antipathy that we see in Urban's volte-face and his need to punish and even humiliate Galileo that lasted really until the end of his life. Urban's extreme language to the Florentine ambassador on how Galileo had transgressed (e.g., "he told me that even our Galilei had dared entering where he should not have, into the most serious and dangerous subjects which could be stirred up at this time", "dealing with the most perverse subject one could ever come across", "matters, involving great harm to religion, indeed the worst ever conceived") is the language of a man who believed something essential was being attacked, not the language of someone who only thought Copernicus "rash" or who was only suffering from hurt pride. His later harping to the Florentine ambassador about Galileo's indifference to Urban's divine omnipotence argument and his forcing necessity on God is, again, a good indicator of his real concerns.

Because of the stringent preventive measures due to the plague, copies reached Rome rather late, two by the end of May and six more in July. Castelli received an early copy and praised it to the skies. But in July the pope was already showing signs of unhappiness about the

---

[23] Galilei, *Dialogue*, 113–14.

[24] Maurice A. Finocchiaro, *The Galileo Affair* (Berkeley: University of California Press, 1989), 222.

book, and he asked Riccardi, without mentioning the pope's name, to have the Florentine inquisitor withhold the book. In August, the Inquisition ordered that the book be suspended, its further sale halted, and its unsold copies confiscated; a special commission was formed to investigate whether Galileo's book had defended heliocentrism, violating the decree of the Index of 1616. This commission most likely consisted of Riccardi, Agostino Oreggi, the pope's personal theologian, and the Jesuit Melchior Inchofer (1585–1649).

Agostino Oreggi (1577–1635), born in Santa Sofia in Tuscany of a modest family, studied at the Roman College, where he earned doctorates in philosophy and theology, and at the University of Rome, where he earned a doctorate in canon and civil law. A protégé and intimate friend of Cardinal Bellarmine, he taught theology in the city of Faenza for nine years and attached himself to Maffeo Cardinal Barberini when he was papal legate at Bologna, becoming his personal theologian. He followed Barberini to Rome as pope, remaining his personal theologian as well as becoming a canon of St. Peter's Basilica and a consultor to the Inquisition and the Congregation of Rites. Between 1629 and 1633 he wrote a slew of theological works: *De Deo uno*, *De individuo Sanctissimae Trinitatis mysterio*, *De angelis*, *De opere sex dierum*, and *De sacrosancto incarnationis mysterio*.

Inchofer was a Hungarian convert from Lutheranism who had joined the Jesuits in 1607 and taught mathematics, natural philosophy, and theology in their college at Messina in Sicily. He was an odd choice insofar as he had come to the attention of Riccardi when a work of his defending the authenticity of a letter supposedly written by the Virgin Mary to the people of Messina was investigated by the Index. He was forced to revise it, and, in fact, just a few weeks before his own highly negative report on Galileo was delivered, an earlier version of the work was put on the Index.

None of these men had a background in science except a bit for Inchofer, but that was not really significant since they were not being asked to judge a scientific question but a theological and legal one: Did Galileo act against the decree of 1616?

In a September 5 letter to the Tuscan state secretary Andrea Cioli (1573–1641), Niccolini described a very emotional meeting he had had with the pope about Galileo the day before, and with Cioli, he now believed that things had taken a disastrous turn. "While we were discussing those delicate subjects of the Holy Office, His Holiness exploded into great anger, and suddenly he told me that even our Galilei had dared entering where he should not have, into the most serious and dangerous subjects which could be stirred up at this time." When Niccolini replied that Galileo had published with official approval, and that Niccolini had himself sent the prefaces to Rome, he received a strong response. "He answered, with the same outburst of rage, that he had been deceived by Galileo and Ciampoli, that in particular Ciampoli had dared tell him that Mr. Galilei was ready to do all His Holiness ordered and that everything was fine, and that this was what he had been told, without having ever seen or read the work." The pope also complained about how Riccardi had been deceived "by having his written endorsement of the book pulled out of his hands with beautiful words, by the book being then printed in Florence on the basis of other endorsements but without complying with the form given to the Inquisitor, and by having his name printed in the book's list of imprimaturs even though he has no jurisdiction over publications in other cities." Niccolini suggested that it would be better just to deal with Cardinal Francesco Barberini, "for when His Holiness gets something into his head, that is the end of the matter, especially if one is opposing, threatening, or defying him, since then he hardens and shows no respect to anyone." He thought it would be best just to temporize and win him over "with persistent, skillful, and quiet diplomacy," with the assistance of his ministers.[25]

Niccolini tried to convince Pope Urban to allow Galileo to know what difficulties and objections were being raised against his work so that he could justify himself, reminding the pope that the book was dedicated to the Grand Duke and Galileo was in the Grand Duke's employ—and that he should take that into consideration. Urban violently replied that that was not the way the Inquisition worked and that Galileo knew full well what the objections were since he had discussed them with him. Urban also stated that even he had prohibited works dedicated to him and that the Grand Duke as a Christian prince should not involve himself since Galileo's doctrine was "extremely perverse", "involving great harm to religion (indeed the worst ever conceived)".[26] The special commission would carefully

[25] Ibid., 229.
[26] Ibid., 230.

go through the book, word for word, "since one is dealing with the most perverse subject one could ever come across".[27] In the same letter, Niccolini recounted his visit to Riccardi, the Master of the Sacred Palace, who complained that the directions for the form of the book given in his letter to the Florentine inquisitor had not been observed, with the preface of a different type and not linked to the rest of the text, and the ending not corresponding to the beginning at all, but he seemed to believe that the book would not be prohibited but only corrected and emended in some points.

In a letter of September 11, Niccolini described a meeting with Riccardi where Riccardi told him that it would be best if the Grand Duke should not press forward with complaints in Galileo's case and with any direct confrontation (that Galileo would benefit from temporizing, proceeding slowly and without noise) since "the Pope believes that the Faith is facing many dangers and that we are not dealing with mathematical subjects here but with Holy Scripture, religion and Faith."[28] He said that Galileo had not followed his instructions and that if the printing had been done in Rome with the page-by-page revision that had been agreed to, it would have been fine. He told Niccolini that he was preparing a revised work that he could then present to the pope to be published and then that would be a good time to write to the pope. He also told Niccolini that the people whom the Grand Duke had proposed to join the special commission to review the work since he did not think the present group sufficiently neutral, Tommaso Campanella and Benedetto Castelli, were not acceptable and that the present group all had a good will toward Galileo and that the Jesuit was chosen by him and was a confidant of his. Finally, he told Niccolini that they found in the files of the Inquisition something which alone was sufficient to "ruin" Galileo completely, the special injunction given by Seghizzi in 1616 not "to hold, teach, or defend it in any way, either verbally or in writing", the Copernican theory which he had promised to obey.

Galileo was in a much weaker position than he had been in 1615–1616 since the Grand Duke himself was in a weaker position. Part of that was due to the relative decline of Italy with the shifting of power and wealth to the north and toward the Atlantic; part was

---

[27] Ibid.
[28] Ibid., 323.

due to the perceived weakness of the reigning Grand Duke; and part was due to the fact that the resources and military forces of the Grand Duchy were already committed elsewhere and could not be readily deployed to overawe or influence its ecclesiastical neighbor. When Cosimo II died in 1621, he left a very young Ferdinando II (1621–1670) under the long regency of his mother, the Archduchess Maria Maddalena, and his grandmother, the Grand Duchess Christina, whose influence continued long after his majority. While the Medici preferred to be relatively autonomous between the two great powers of France and Spain (witness Fernando I's extensive financial and diplomatic support for Henry of Navarre's pursuit of the French throne, the marriage of his niece to him, and his own marriage to the French Christina of Lorraine), nonetheless by 1632 they were again part of the Habsburg orbit. In 1619, early in the Thirty Years' War, five hundred Tuscan cavalrymen were sent to Bohemia, and from 1632 to 1639 Ferdinando II's two brothers Mattias and Francesco fought with a Tuscan contingent in Germany. (Francisco died of the plague there in 1634.) These were all extremely costly ventures, and the military budget of Tuscany from 1625 to 1650 was half of the Grand Duchy's income.

There were many reasons for their support for Spain and the Habsburgs. Beyond the family connections by marriage (the present Grand Duke's mother was an Austrian Habsburg, and his sister would marry a Habsburg archduke) and the military obligations due to receiving the fief of Siena which we have already seen, there were clear financial and practical reasons connected to the strong trading links between Tuscany and the Spanish realm and the large number of Tuscans serving in the Spanish military. There was also the fact that Spain owed massive sums to the Medici. Finally, Spanish control of Italy and especially of Milan directly to their north had made things far more peaceful, preserving Tuscany from invasion and the costs that followed. It was while the Tuscan forces were busy in Milan and the plague raged in Tuscany that Pope Urban seized the Duchy of Urbino in 1631 after the death of Francesco della Rovere II, the last male of the line, denying the claims of Grand Duke Ferdinando II, who had arranged to be married to Francesco's sole heir, his granddaughter Vittoria della Rovere, with the primary intention of gaining the duchy. The Medici both as a Habsburg ally and as a claimant to

Urbino would have a lot less pull in the Rome of Pope Urban VIII than in previous pontificates.[29]

In September the special commission finished its report. A number of things were found: that Galileo had used the Roman imprimatur without permission and improperly; that the preface which contained his support for the decision of the anti-Copernican decree of 1616 was in different type and rendered useless by its separation from the body of the work; that he put the "medicine of the end" (Urban's divine omnipotence argument) "in the mouth of a fool and in a place where it can only be found with difficulty, and then he had it approved coldly by the other speaker by merely mentioning but not elaborating the positive things he seems to utter against his will"; "that many times in the work there is a lack of and deviation from hypothesis, either by asserting absolutely the earth's motion and the sun's immobility, or by characterizing the supporting arguments as demonstrative and necessary, or by treating the negative side as impossible"; that he "treats the issue as undecided and as if one should await rather than presuppose the resolution"; that he mistreated the contrary authors and those most used by the Church; that he wrongly asserted a certain equality between the divine and human intellects; that he gave as an argument that supporters of Ptolemy occasionally become Copernicans but never the reverse; that he wrongly attributed the ebb and flow of the tides to the motion of the earth. Finally the report mentioned the injunction that Galileo had received and accepted to abandon completely the opinion that the sun was the center of the world and the earth moves, which he could not "henceforth hold, teach, or defend in any way whatever, orally or in writing".[30]

It was decided that the book would be brought before the Inquisition, and on September 15, "as a favor to the Grand Duke", Niccolini was told of this. Niccolini's protests, then and a few days later to the pope himself, were of no avail. The pope told him that while Galileo was a friend and as mathematician to the Grand Duke was salaried and employed by him, and known as such, his opinions had

[29] Niccolo Capponi, "Le Palle di Marte: Military Strategy and Diplomacy in the Grand Duchy of Tuscany under Ferdinand II de' Medici (1621–1670)", *Journal of Military History* 68 (2004): 1105–41.

[30] Finocchiaro, *Galileo Affair*, 218–22.

been condemned in 1616. These opinions were dangerous and pernicious, and "Galileo had gotten himself into a fix which he could have avoided."[31] On October 1, Galileo was personally ordered to go to Rome, summoned by the Inquisition, and despite various attempts to delay the trip, or not to go and respond to objections in writing or to appear before someone else, his old age and ill-health being major impediments to travel, these were rejected.

In a January 15, 1633, letter to Diodati, Galileo talked about how great an abuse it was to use Scripture on questions dealing with natural phenomena and that it should be prohibited. He also told him that he had been summoned to Rome by the Inquisition. "From reliable sources I hear the Jesuit Fathers have managed to convince some very important persons that my book is execrable and more harmful to Holy Church than the writings of Luther and Calvin." He was sure it would be prohibited though he had gone personally to the Master of the Sacred Palace who had examined "very minutely" each page, and having licensed it, had it reviewed again in Florence by another who changed a few things, asked to be excused, and predicted that Galileo "would be dealing with very bitter enemies and very angry persecutors, as indeed it followed". Galileo feared that he would never be able to finish his other works, especially the one on motion.[32]

Galileo left Florence by litter for Rome on January 20, 1633, though with the two weeks he spent near Acquapendente due to the plague quarantine, he arrived at Rome only on February 13. After an initial flurry of visits to those who could help his cause, Galileo was advised that he should neither receive nor pay visits during his stay, and he was allowed to remain in the Tuscan embassy instead of in the prison of the Inquisition as a favor to the Grand Duke. He waited two months before he was called by the Inquisition for interrogation, arriving there on the morning of April 12, 1633. All during this time Niccolini was doing all he could do, visiting powerful cardinals and meeting with Pope Urban, pleading Galileo's advanced age, ill-health, and docility as reasons for clemency, the Grand Duke also writing letters to all the Inquisition cardinals.

---

[31] Ibid., 234–37.
[32] Ibid., 223–26.

Niccolini's interviews with Pope Urban did not bode well for Galileo. Urban's anger had not been assuaged, and he was told that the judicial process would take its course. In a meeting on February 26 Urban complained that Galileo should never have published his opinions for which Ciampoli was to blame, that Galileo claimed he only wanted to discuss the earth's motion hypothetically while in fact he did so assertively and conclusively, that he had disobeyed the injunction of 1616, and that his doctrine was bad. In a meeting on March 13 Urban again criticized Ciampoli and Galileo for meddling where they should not, and that while Galileo had been his friend, and conversed and dined with him familiarly, the interests of faith and religion were preeminent. When Niccolini told him that he was sure that if Galileo were heard, he would give every satisfaction, Urban replied that Galileo would be examined in due course, "but there is an argument which no one has ever been able to answer: that is, God is omnipotent and can do anything; but if He is omnipotent, why do we want to bind him?" Niccolini attempts to reason with the pope were of no avail.

> I said that I was not competent to discuss these subjects, but I had heard Mr. Galilei himself say that first he did not hold the opinion of the earth's motion as true and then that since God could make the world in innumerable ways, one could not deny that He might have made it this way. However, he got upset and told me that one must not impose necessity on the blessed God; seeing that he was losing his temper, I did not want to continue discussing what I did not understand, and thus displease him, to the detriment of Mr. Galilei.[33]

Galileo was certainly under great strain, and his infirmities only got worse. In a letter to the Cioli (April 8, 1632), Niccolini recounted that Galileo had spent two whole nights moaning and screaming in arthritic pain. Galileo was even more distressed by Niccolini's advice that, for a quick resolution, he should just stop defending his opinions and submit to what they wanted him to believe. He appeared so depressed that Niccolini feared greatly for his life.[34]

[33] Ibid., 247.
[34] Ibid., 249.

In 1628 Pope Urban had ordered that the Inquisition would no longer meet at the palace of its senior member but in the apartments of the Dominican Master General at Santa Maria sopra Minerva. So when Galileo arrived on April 12, for his first interrogation, it was to Santa Maria sopra Minerva. He came not before its assembled cardinals in full panoply, but before just two figures, the Commissary General Vincenzo Maculano (1578–1666) and his assistant, the Prosecutor of the Tribunal, Carlo Sincero, both Dominicans. We tend to have this image of a trial before the Inquisition as some grand meeting of cardinals and officials in their official splendor, with a procession of witnesses, of the defenders and the accused as in a modern court case, and with all the drama that that implies. But in fact the activities of the Inquisition were rather bureaucratic and very undramatic, with all the drama kept for the solemn abjuration at the end. The Inquisition itself was one of the fifteen congregations of the Curia, a committee of about ten cardinals assisted by four principal officials but with the difference that its sole head was the pope. It usually met twice a week, on Wednesdays without the pope at Santa Maria sopra Minerva and on Thursdays with him either at the Vatican or the Quirinal Palaces. They were assisted by a body of theological consultors who met on Mondays.

At this time the cardinals on the Inquisition were Gaspar Borja, Felice Centini, Guido Bentivoglio, Desiderio Scaglia, Laudivio Zacchia, Berlinghiero Gessi, Fabrizio Verospi, Marzio Ginetti, Francesco Barberini, and Antonio Barberini Sr. Except for Borja, Centini, and Scaglia, the Inquisition was totally dominated by loyalists of the Barberini and would produce whatever result the pope desired. Borja had been a cardinal since 1611, becoming a member of the Inquisition since 1617, which he only attended sporadically, and, as we already saw, was an implacable enemy of Pope Urban VIII and would eventually be forced out of Rome. Centini (1562–1641), born of poor parents in the Marches of the Papal States, favored by Sixtus V and by his nephew Cardinal Montalto (who were also from the Marches), rose through the Conventual Franciscans, becoming a consultor to the Inquisition, procurator general of his order in 1609, and finally a cardinal and member of the Inquisition in 1611 under Paul V. According to an Italian spy of the English king James I, in the summer of 1614 he had been caught at the house of his prostitute to the greatest scandal of all.[35] Because of his nephew's trial for attempting to kill the pope with black magic, which we have already mentioned, he was in exile at Macerata, his episcopal see in the Marches, until the final phase of Galileo's trial. Scaglia (1567–1639) was from Brescia, though he claimed Cremona as his birthplace since Brescia was in Venetian territory and Venice was not popular in the Rome of Paul V. Of lowly origins, he became a Dominican and served as inquisitor at Pavia, Cremona, and Milan before being made Commissary General of the Holy Office from 1616 to 1621, and a cardinal in 1621. Pro-Spanish and ambitious for himself, he had opposed the election of Urban VIII, earning the hostility of the Barberini, though he seems to have made up with them by 1632. He acquired a substantial fortune and art collection, and lived in several impressive palaces. The most experienced of inquisitors, he was the author not only of a practical inquisitorial manual in Italian which was widely circulated in manuscript (*Prattica per procedere nelle cause del S. Offizio,* c. 1635) but also of an important *Instructio* which condemned the abuses practiced in pursuing witches.[36]

It is worth looking at the remaining members. There was, of course, Antonio Barberini Sr., the pope's brother who was also the Secretary of the Inquisition, and Francesco Barberini, the pope's nephew and Secretary of State. Bentivoglio (1577–1644) was born in Ferrara, where his brother Ippolito had opposed militarily the papal annexation to Guido's great discomfort for fear of his own advancement and for which he apologized. A descendent of the Bentivoglio of Bologna who had ruled that city in the fifteenth and early sixteenth centuries, he was educated at Ferrara and Padua, where he received a doctorate in both canon and civil law, and where he met Galileo, who taught him mathematics. He served Paul V as papal ambassador to the court of the Archdukes Albert and Isabella in the Spanish Netherlands from 1607 to 1615 and then in Paris to the French king from 1615 to 1621, being made a cardinal by Paul V in 1621. An able diplomat and skilled historian, he was pro-French and also an intimate of Pope Urban VIII. Verospi (1571–1639) was born in Rome,

---

[35] Mormando, *Bernini,* 21.

[36] John Tedeschi, "The Roman Inquisition and Witchcraft: An Early Seventeenth-Century 'Instruction' on Correct Trial Procedure", *Revue de l'histoire des religions* 200 (1983): 163–88.

of a noble family originally from Spain, studied at the Roman College and then law at Rome, Ferrara, and Bologna, earning a doctorate in canon and civil law at the last, and spent the rest of his career working up the curial ladder under the patronage of the Aldobrandini, serving in the Curia itself, especially as an auditor (judge) on the Rota, the highest appeals court in the Catholic Church, on diplomatic missions and as a governor in the Papal States. He became very close to Carlo Barberini, the pope's brother, successfully arranging the marriage of his son Taddeo Barberini to Anna Colonna and being made a cardinal in 1627 under Pope Urban. Berlinghiero Gessi (1563–1639), of a patrician family and a distant relative to Pope Gregory XIII, was born in Bologna, where he received a doctorate in canon and civil law, went to Rome to assist his uncle who was an auditor on the Rota, and practiced law there before returning to teach law in Bologna. He served in various ecclesiastical positions before becoming vicar-general of Rome in 1599 and vicegerent of Rome from 1600 to 1607. He was ambassador to Venice from 1607 to 1618, governor of Rome from 1618 to 1623, prefect of the Apostolic Palace (papal household) in 1622 and under Pope Urban until 1625. He was governor of newly annexed Urbino from December 1624 until May 1627, and was made a cardinal in 1626 by Urban, though he only joined the Inquisition in 1629. Marzio Ginetti (1585–1671) was born in Velletri near Rome of a noble family from Bergamo, though many sources incorrectly claim him to be of humble origins. He was educated by the Jesuits in Rome and received a doctorate in canon and civil law. Patronized by Paul V and Cardinal Odoardo Farnese, he worked his way up in the Roman Curia, replacing Gessi as prefect of the Apostolic Palace and pontifical household in 1626. Made a cardinal *in pectore* (in secret) in 1626 by Urban, which was made public in 1627, he joined the Inquisition in 1629. A favorite of Urban VIII and his brothers, the Venetian ambassador wrote that Urban considered him his fourth nephew.[37] Zacchia (1565–1637), brother of Paolo Emilio Cardinal Zacchia, was born at the castle of Vezzano in the Republic of Genoa, gained a doctorate in canon and civil law at Pisa, married, had two children, and then joined the clergy after the death of his wife. He worked in Rome

[37] Thomas F. Mayer, *The Roman Inquisition: A Papal Bureaucracy and Its Laws in the Age of Galileo* (Philadelphia: University of Pennsylvania Press, 2013), 87.

advancing under the patronage of the Aldobrandini. In 1605 he was made bishop of Montefiascone (succeeding his brother Paolo Emilio and resigning the diocese in 1630 to his nephew), served in various positions including as ambassador to Venice under Pope Gregory XV, and was made a cardinal in 1626 by Pope Urban.

Vincenzo Maculano seemed an odd choice for the office of Commissary General. Born in the Duchy of Parma and joining the Dominicans at sixteen, he had studied at Bologna not only law and theology but also mathematics and geometry, and owed his rise to his ability as a military engineer and to the patronage of the Barberini, taking his oath as Commissary General on December 22, 1632. He replaced Ippolito Lanci (c. 1571–1634), a pro-Spanish dependent of the Ludovisi who had told Castelli in October 1632 that he meant to write in defense of Copernicanism and that he did not think the issue should be decided on the basis of Scripture.[38] While Maculano remained in that office until 1639, when he became Master of the Sacred Palace at the death of Riccardi, he continued his work as a military engineer, working in Piedmont in 1636 and working on the fortifications of Malta in 1638. Even after he was made a cardinal in 1641, he continued working on fortifications. While not the most acute theologian, he was perhaps the kind of man who might understand and sympathize with Galileo better than many. In any case, as a creature of the Barberini, he would execute their desires to the letter.[39]

Galileo was interrogated in Latin, and his replies were recorded in Italian. After a number of introductory questions, he was shown a copy of the *Dialogue*, and after he had confirmed that it was his and where and when he had composed it, he was asked about what happened in his 1616 visit to Rome, the decision to condemn Copernicus, and of his meeting with Cardinal Bellarmine. He replied that Cardinal Bellarmine had told him that Copernicus' opinion could neither be held nor defended, and he presented a copy of the certificate given him by Bellarmine to prove it. While he did remember some Dominicans being present, he did not remember if they were there already or came later. Nor did he remember any other sort of injunction except the oral one of Bellarmine's not to hold

[38] Ibid., 125.
[39] Ibid., 11, 76, 125–29.

or defend Copernicanism. He did not remember the phrases "not to teach" and "in any way whatever", though that might be, he thought, because they were not in Bellarmine's certificate, the thing by which he remembered the events. When asked about whether he told Riccardi about the injunction, he replied that he did not since he was not attempting to hold, defend, or teach that opinion, but was trying to refute it, showing that Copernicus' reasons were invalid and inclusive.

Instead of being placed in the cells of the Inquisition, Galileo was given the prosecuter's suite of rooms, and his servant was allowed to stay and assist him, his meals coming from Niccolini's kitchen. On April 17, the reports of the three members of the second special commission (Oreggi, Inchofer, and the Theatine theologian Zaccaria Pasqualigo) on whether the *Dialogue* defended Copernicanism came in, and all three consultants agreed that it did, with Inchofer's report being the longest and most critical of Galileo. On April 21 the congregation accepted the judgment of the reports that Galileo had indeed violated the Inquisition's special injunction. Seeing by this decision that the case was now for all intents and purposes over and continued resistance by Galileo would only necessitate the ominous-sounding "greater rigor in the administration of justice" (often a code word for torture), and seeing that the continuation of the trial could help no one, Galileo least of all, Maculano received permission to deal with him extra-judicially. On April 27, after much argument, he made Galileo realize his error and confess that he had gone too far in his book, though Galileo wanted some time to find the right words. So in his second deposition of April 30, he stated that after much thought, and after rereading the *Dialogue*, which he had not looked over in the last three years, he realized that he had indeed, without his intention, done exactly as the Inquisition had charged him: that despite his intention to confute these arguments, he had in fact made them stronger and more convincing than they really were. He did it, however, out of a desire not to make the opposing arguments too weak and animated by the natural gratification everyone feels to show off his subtlety and ingenuity by finding probable reasons for false propositions. His error was "one of vain ambition, pure ignorance and inadvertence", and he even offered to add one or two days to the *Dialogue* to prove Copernicanism more clearly false. Once he had confessed he was allowed to return to the Tuscan embassy, where he would await the final summons of the Inquisition.

Various reasons have been given for this change of mind on Galileo's part. Blackwell sees it as part of a plea bargain where Galileo would admit to some portion of the crime for a reduced sentence, a plea bargain that went awry in the end.[40] Wootton, utilizing a recently discovered (1999) unsigned and undated document in Inchofer's handwriting attacking Galileo for atomism in his *Assayer*, believes that Galileo capitulated because Maculano threatened him with the possibility that he would bring up the far more damaging charge of denying the Catholic doctrine of Transubstantiation if he did not confess.[41] Both hypotheses are very plausible, but both are unnecessary. At this point Galileo had already lost his case, and he would only have to be made aware of that fact and its possible consequences for him to see what he had to do to save himself. While there may have also been the suggestion of torture, Galileo's age and ill-health would have made its application unlikely. In any case, even a man as proud and stubborn as Galileo would have eventually seen the writing on the wall, and, considering the terrible distress he had been going through, he could have seen it the wisest and quickest way out of his difficulties.

Certainly there were many who wanted to minimize the difficulties on all sides. A confession on Galileo's part would also obviate the difficulties of explaining the two opposing documents concerning the special injunction and the fact that a book with not one but two imprimaturs was being condemned. It would lower the case's profile, lesson the tension, and save face for all. As Maculano stated in his April 28 letter to Cardinal Barberini: "The Tribunal will maintain its reputation; the culprit can be treated with benignity; and, whatever the final outcome, he will know the favor done to him, with all the consequent satisfaction one wants in this."[42] There was also a deep

[40] Richard J. Blackwell, *Behind the Scenes at Galileo's Trial* (Notre Dame, Ind.: Notre Dame University Press, 2006), 13–26.

[41] David Wootton, *Galileo: Watcher of the Skies* (New Haven, Conn.: Yale University Press, 2010), 221–25; and Mariano Artigas, William Shea, and Rafael Martinez, "New Light on the Galileo Affair?", in *The Church and Galileo*, ed. Ernan McMullin (Notre Dame, Ind.: Notre Dame University Press, 2005), 213–33.

[42] Finocchiaro, *Galileo Affair*, 277.

concern over Galileo's health as a letter of April 22 from Maculano to Francesco Cardinal Barberini, only discovered in 1998 and published in 2001, reveals:

> Last night Galileo was afflicted with pains which assaulted him, and he cried out again this morning. I have visited him twice, and he has received more medicine. This makes me think that his case should be expedited very quickly, and I truly think that this should happen in light of the grave condition of this man. Already yesterday the Congregation decided on his book and it was determined that in it he defends and teaches the opinion which is rejected and condemned by the Church, and that the author also makes himself suspected of holding it. This being so, the case could immediately be brought to a prompt settlement, which I expect is your feeling in obedience to the Pope.[43]

After Galileo returned to the embassy on April 30, things seemed to go much better. His health returned, and the trial seemed well on the way to being wrapped up. On May 10, Galileo returned to the Inquisition where he left a written defense and the original of Cardinal Bellarmine's certificate, and asked for "the usual mercy and clemency of the tribunal". On May 21, Niccolini spoke with the pope and Cardinal Francesco Barberini and was told that the trial could easily be settled at the second meeting of the Inquisition the Thursday of the next week (June 2). Niccolini could very well imagine that the book would be prohibited, though they might decide that it just needed to include an apology by Galileo as he had suggested. Galileo would also be given some penance for disobeying the precept given him by Bellarmine in 1616. In May or June the summary report on the trial, from which the final decision would be made, was sent to the pope and the cardinals. Unfortunately, it was a very biased and highly negative document that could do Galileo no good. If Urban had had any thoughts about leniency, they seemed to have completely disappeared, for at the June 16 meeting of the Inquisition, with the pope presiding, it was decided that the book would be condemned, Galileo imprisoned, and some sort of penance imposed upon him. He

---

[43] Blackwell, *Behind the Scenes at Galileo's Trial*, 14.

would also be summoned to find out his true purpose in writing the *Dialogue*, even using torture if necessary. On June 21 Galileo returned to the palace of the Inquisition, where he repeated, again and again, that after the decision of 1616 he had not held to Copernicanism, even when he wrote the *Dialogue*, even after he was threatened with torture if he did not tell the truth. He signed his deposition and left.

At the June 22 meeting of the Inquisition at the monastery of Santa Maria sopra Minerva, with three cardinals absent (Borja, Zacchia, and Francesco Barberini), but surrounded by twenty other witnesses, Galileo, kneeling and dressed in the white robes of a penitent, heard his sentence read out. After narrating the history of the case from 1615 on, they gave their final verdict: that Galileo was vehemently suspect of heresy for supporting that the sun was the center of the world and does not move from east to west, that the earth moves and is not the center of the world, and that one may hold and defend as probable this opinion even after it has been declared as contrary to Scripture.

The *Dialogue* would be prohibited, Galileo would receive a "salutary penance" of reciting the seven penitential psalms once a week for three years, and he would be formally imprisoned at the pleasure of the Inquisition. Then, still kneeling and with a lighted candle in his hand, Galileo read out his abjuration where he utterly rejected ("abjure, curse and detest") the heresies and errors of Copernicanism and swore never again to say, assert, or write anything that might cause a similar suspicion. He then signed it, and on the next day his imprisonment was reduced to house arrest at the Villa Medici.

There were Jesuits who still held an affection for Galileo, such as Grienberger, and those who certainly hated him or held mixed feelings for him, but except for Inchofer no Jesuits were involved in the trial of 1633. And there was certainly no Jesuit conspiracy. During the trial Inchofer wrote a lengthy scriptural and theological treatise against heliocentrism, entitled *Tractatus syllepticus*, which was published in 1633, but he would eventually (in 1648) be put on trial by his own order, the Jesuits, and punished by them.[44] Christopher Scheiner, whose *Rosa ursina* (1626–1630), his definitive study of sunspots, actually may have aided Galileo in arguing for his

[44] Ibid., 44.

Copernicanism, and who disliked Galileo strongly, was in Rome from 1624 to 1633, working at the Roman College, but he was not involved in any aspect of the affair. He did write an attack on the *Dialogue* entitled *Prodromus pro sole mobili et terra stabili contra Academicum Florentinum Galilaeum a Galilaeis* that was finished a month after Galileo's condemnation, but even here it was focused almost completely on sunspots—which are not the most important part of Galileo's book. For all his dislike of Galileo, he may have been a secret Copernican, and his book was denied for publication by his Jesuit superiors, not appearing until 1651.[45]

On July 30 Galileo was allowed to stay under house arrest with his friend Archbishop Ascanio Piccolomini of Siena, who treated him as an honored guest for six months. His health improved, and he began again his studies. In December he was allowed to return to his villa at Arcetri, but he was not allowed to have guests except for family and friends, nor have large gatherings, nor go to Florence. His home was close to his daughters' convent, the elder, Suor Maria Celeste, being his greatest comfort in his old age. But she died on April 1, 1634, shortly after he returned, leaving him in extreme grief. Except for part of 1638 when he stayed in Florence, he remained in Arcetri with limited access to visitors. There he completed his *Discourse on the Two New Sciences*, a dialogue with the same characters as his earlier dialogue, which was published in 1638 by the highly respected press of Louis Elzevier at Leiden in Protestant Holland. (It should be noted that it did receive an imprimatur to be printed in Olmütz in Habsburg Moravia and in Vienna, and by a Jesuit no less, but other difficulties and the clear superiority of Elzevier won out in the end.) Galileo's *Starry Messenger* and his crusade for Copernicanism had distracted him from finishing this his greatest work, the work by which he would change the face of physics and for which so many consider him the father of modern science. Also, as Finocchiaro has pointed out, although the forbidden topic of the earth's motion was not mentioned in the work, the laws of motion elaborated there would help provide later, through the work of Isaac Newton (1642–1727) and his *Principia Mathematica* (1687), a more effective proof for it, at least at the theoretical level, than he could. "Here is one of those ironies

45 Ibid., 65–91.

of history, where the temporary victor of a particular battle creates conditions that pave the way for his eventually losing the war."[46]

In 1636 his eyes began to deteriorate, and in 1637 his sight declined so rapidly that he was totally blind by the end of the year. As he wrote to Diodati on January 2, 1638:

> Alas, My Lord, Galileo your dear friend and servant has been for a month totally and incurably blind. Think, my Lord of my afflictions when I consider that sky, that world, that Universe, which I, with my remarkable observations and clear demonstrations have enlarged a hundred and a thousandfold, beyond what was universally seen by the learned of all past centuries, are now so shrunk and limited to me that they are no greater than the space occupied by my person.[47]

While he had various assistants stay with him, such as Vincenzio Viviani and Evangelista Torricelli, despite pleas from various sources (he was told in 1634 to stop petitioning for a pardon or he might be sent to a real prison in Rome), he remained under house arrest until his death. On January 8, 1642, late in the evening, with his son Vincenzio, Vincenzo Viviani, and Evangelista Torricelli at his side, Galileo breathed his last. The Grand Duke wished for a splendid tomb for him in Santa Croce, the great Franciscan church of Florence where so many of its most famous citizens are buried, but the Holy See denied his request, saying that it would be scandalous for one whose doctrine had been condemned and who had died under penance to be buried there. So his body remained in a room behind the sacristy for almost a hundred years until 1737 (during the reign of the Florentine pope Clement XII), when it was buried with much ceremony and with all of Florence's nobility, religious leaders, and learned men attending, in an impressive mausoleum in the body of the church in front of the tombs of Michelangelo and Machiavelli with money left for the purpose by Viviani.

While the story of Galileo's life quieted down, the story of the Galileo Affair took on a new intensity and a life of its own, as is well told in Maurice Finocchiaro's *Retrying Galileo 1633–1999* (2005). To

[46] Galileo Galilei, *On the World Systems: New Abridged Translation and Guide*, trans. and ed. Maurice Finocchiaro (Berkeley: University of California Press, 1997), 47.

[47] Giogio Abetti, "Galileo the Astronomer", *Popular Astronomy* 159 (1951): 143.

a great extent this was due to Pope Urban himself, who wanted Galileo's humiliation to be complete and to generalize the condemned opinion. He had copies of Galileo's sentence and abjuration sent to all the papal nuncios and local inquisitors to be publicized, something never done before or after. They were particularly to make it known to all professors of philosophy and mathematics, and even posters and flyers were printed to make the public aware of it. It was after seeing one of these posters that Descartes, then living in Holland, decided not to publish his *Le Monde* for fear of criticism or condemnation. By and large, the Catholic states outside Italy (with Venice the exception within Italy) did not react well or even accept the condemnation, though for various reasons; and they tended to see it as an abuse of power and did not cooperate to enforce it.[48] It had no legal standing in France, where it was not accepted by the Sorbonne and the Parlement of Paris, or in Spain, where the 1634 Index decree condemning the *Dialogue* was not included in the Spanish Index, just as Copernicus' *De revolutionibus* was not on its Index.

With time and with greater discoveries such as the 1728 discovery of aberration of starlight by the English astronomer James Bardley, the opposition to Galileo and Copernicus weakened. In 1744, during the pontificate of the liberal and highly learned Prospero Lambertini, Benedict XIV (whom even Voltaire had kind words for), and with the approval of the Inquisition, the first complete edition of Galileo's works was published in Italy by the seminary press in Padua. It included the banned *Dialogue* (with the marginal postils removed or corrected) but not the *Letter to the Grand Duchess Christina*. It included Galileo's sentence of condemnation and abjuration, a preface which affirmed the purely hypothetical and mathematical (not physical) nature of the heliocentric theory, and a dissertation on biblical interpretation by the respected Benedictine biblical scholar Dom Augustin Calmet (1672–1757). Calmet's essay was rather pro-Galilean insofar as it talks about how God adapts his revelation to the simple and their way of speaking and how we should not look there for philosophical rigor or the precision of the human sciences. Also during Pope Benedict XIV's reign the Index of 1758 removed the general prohibition

<hr />

[48] Finocchiaro, *Retrying Galileo*, 72.

on books that taught the mobility of the earth and the immobility of the sun, though it did not remove the previously condemned books of Copernicus, Foscarini, Galileo, Kepler (whose *Epitome of Copernican Astronomy* had been condemned in 1619), and Zuñiga. Only in the Index of 1835 were these five books finally removed.

But it is one thing for the Catholic Church to remove Galileo's book from the Index and another to admit that any sort of mistake had been made in the first place, and there was a great deal of resistance to the latter. In 1941, to mark the tricentennial of Galileo's death the Pontifical Academy of Sciences commissioned Monsignor Pio Paschini (1878–1962), rector of Pontifical Lateran University and a respected historian, to write a biography of Galileo. When he finished the manuscript and delivered it in 1945 to Church authorities, it was rejected for publication, with the objections of the Holy Office being the most weighty. It was seen as an apology for Galileo, as too critical of the Jesuits, and as admitting the Church had made a mistake in condemning Galileo even by the standards of the day. He refused to change his judgments substantially, and so the book languished in a kind of limbo. When it was printed in 1964, two years after Paschini's death, it was heavily edited and revised by a Belgian Jesuit, so that in many places it had partially or completely altered what he thought and in a few places says the exact opposite of what he had written.[49]

More recently, under Pope John Paul II (r. 1978–2005), another somewhat more successful attempt has been made to rehabilitate Galileo and admit that mistakes were made by the Church.[50] In 1979, at a commemoration of the centennial of Einstein's birth, in a speech at the Pontifical Academy of Sciences, Pope John Paul II stated that Galileo "had to suffer a great deal at the hands of men and organisms of the Church",[51] implying that Galileo's treatment was not just an error but an injustice. He later created a study commission to investigate the Galileo Affair, and while there were a number of excellent publications that came out of this revived interest, particularly

---

[49] Ibid., 318–37.

[50] Ibid., 338–57; and George V. Coyne, "The Church's Most Recent Attempt to Dispel the Galileo Myth", in McMullin, *The Church and Galileo*, 340–59.

[51] Finocchiaro, *Retrying Galileo*, 340.

in connection with the Vatican Observatory, the final result in 1992 was something of a letdown. The report given by Cardinal Poupard, the head of the commission, was something of a regression, for while it acknowledged the theological errors of his judges (but not Bellarmine), it also seemed to blame Galileo for not proving his case. Even in Pope John Paul II's address which followed the reading of the report a certain regression could be noted. He called the Galileo Affair a "tragic mutual incomprehension", and, like Poupard, spoke about how Galileo was a better theologian than his opponents and put the blame on the errors of the majority of theologians, though again not Bellarmine. No mention was made, however, of popes, cardinals, and the other officials of the Church who were involved, so one would suppose there was no blame on them. While newspapers trumpeted the headline that the Vatican had admitted that Galileo was right, the reality was more complex.

While Pope Benedict XVI (r. 2005–2013), as pope at least, has celebrated and praised Galileo a number of times for his extraordinary synthesis of faith and reason and deep awe before the wonder of God's creation, there was only one place, as far as I can tell, where Pope Benedict has admitted the Church's error in the Galileo case: in a talk he gave before the priests of Rome on February 14, 2013, three days after he stunned the world by announcing his resignation of the papacy. He spoke about being a young theologian at the exciting time of the Second Vatican Council (1962–1965).

So we went off to the Council not just with joy but with enthusiasm. There was an incredible sense of expectation. We were hoping that all would be renewed, that there would be a new Pentecost, a new era for the Church.... However, there was a feeling that the Church was not moving forward, that it was declining, that it seemed more a thing of the past and not the herald of the future. And at that moment, we were hoping that this relation would be renewed, that it would change; that the Church might once again be a force for tomorrow and a force for today. And we knew that the relationship between the Church and the modern period, right from the outset, had been slightly fraught, beginning with the Church's error in the case of Galileo Galilei; and we were looking to correct this mistaken start and to rediscover the union between the Church and the best forces of the world, so as to open up humanity's future, to open up true progress.

Thus we were full of hope, full of enthusiasm, and also eager to play our own part in this process.[52]

While the Church has come a long way in recognition of its share of the blame in the Galileo Affair—and probably farther than most institutions—some ambivalence remains, and will probably remain until that new Pentecost happens that Pope Benedict XVI spoke about. In the conclusion we will draw together our thoughts from what we have already seen and attempt some final discussion.

[52] Benedict XVI, "Meeting with the Parish Priests and Clergy of Rome, Feb. 13, 2013", *L'Osservatore Romano*, English edition, February 20, 2013, 8.

CONCLUSION

On December 5, 1634, Nicolas-Claude Fabri de Peiresc (1580–1637), a French nobleman, priest, and one of the great patrons of learning in early seventeenth-century France, wrote to Francesco Cardinal Barberini—whom Peiresc had hosted during Barberini's visit to France in 1625 and with whom he was in frequent correspondence—asking for a mitigation of Galileo's sentence. Peiresc, himself an able scholar and a leader in the Republic of Letters, had studied at Padua (1599–1601), where he had become acquainted with Galileo and had even been tutored by him. He had been appalled by Galileo's humiliation and punishment. Despite the tepid response he received from Barberini, and indifferent to the negative reaction that might follow, he asked again on January 31, 1635, stressing that the punishment of Galileo would be compared by posterity with the persecution of Socrates, the ancient Athenian sage and the father of Western philosophy. He was not successful, but he was the first of many to make this comparison of Galileo with Socrates, for in many ways they were similar.

Like Socrates, Galileo was a charismatic teacher who attracted many enthusiastic disciples and won them over to a wholly new view of the world. Like Socrates, Galileo was an indefatigable seeker of truth, a man possessed with an amazing verbal and dialectical dexterity ready to challenge the complacent with their unreflective conformity, a gadfly ready to puncture the pretensions of the so-called wise and self-satisfied. Like Socrates, Galileo too had a fundamental optimism and belief in reason, a buoyant exuberance and irreverence that could be to some extremely attractive but to others irritating and off-putting. Finally, like Socrates, Galileo felt he was on a divine mission, a mission that he could neither refuse nor moderate. Unlike Socrates, however, he drank no hemlock. And while the sufferings of Galileo had been severe, and should not be forgotten, there is still something ignoble if understandable in his capitulation and perjury, in his denial of the real intentions of his great work and his abject abjuration before the Inquisition. Lacking in Socrates also—at least in the highly stylized dialogues of Plato by which we know him— were Galileo's vanity and inflated ego, his ambition, his need to protect his scientific preeminence and status against all comers, and the self-destructive pettiness and nastiness that mar his otherwise highly admirable personality.

It seems to me that if one had to compare Galileo to anyone in the classical world, it might be, rather, to Oedipus than to Socrates. The legendary figure of Oedipus, made forever memorable through the great plays of Sophocles, was also an extraordinary man. He too was a man of great intelligence, brave and ingenious, as we see in his standing up to the monstrous Sphinx and the solving of its riddle. He too was a man of action, incapable of passivity, unrelenting in his investigations, in his pursuit of truth, even when warned that it would be far better to leave things alone. And like Galileo, his intemperate zeal and pride in his own ability would bring on his destruction, a disaster that need not have happened, leaving him in the end bowed and blind, though unbroken, an object of pity but also of respect, heroic in stature through his tragedy.

J. L. Heilbron, in his recent biography of Galileo, sees him as a quixotic figure (and Galileo did possess an Italian translation of Cervantes' great novel in his library) who was the cause of his own failure. He believes that Galileo experienced a sort of epiphany at forty-five (when he was far beyond the midpoint of the average person's life in those days) under the impetus of his telescopic discoveries, changing him from a relatively quiet and circumspect individual to "a knight errant, quixotic and fearless, like one of the paladins in his favorite poem, Ariosto's *Orlando furioso*", a megalomaniac attacking anyone who opposed his opinions, making his future collision with the pope (the Church), as Heilbron believes, intelligible and even inevitable.[1] You see much the same in Richard Westfall: Galileo, a middle-aged man who up to this time had produced a few small pamphlets, appears suddenly and from nowhere to reveal a new world, an overachiever bent on worldly success, an overzealous Copernican who staked everything on the argument of the tides, who was the architect of his own downfall, and

[1] John L. Heilbron, *Galileo* (Oxford: Oxford University Press, 2010), vii.

brought on the very catastrophe he most feared and most desired to avert.[2]

There is more than a bit of truth in all that, but that being said, there is more than plenty of blame to go around, and Galileo was far from the only culprit. There was the envy and malice of those who felt Galileo had risen too far and too fast: someone who had been peremptorily vaulted from obscurity into preeminence, a celebrity over whom princes and kings and popes fawned over and coddled. Then there were those whose hatred and opposition was based on their vested interest in the old ways or out of their fear of the new—the cross of innovators throughout history. And when Galileo experienced hostility and opposition, especially when he thought it unjustified, it only made him more aggressive.

The Catholic Church, too, must bear a very, very large share of the blame. Since the Reformation it had become much more conservative, much more defensive, bureaucratic, and controlling—and much less able to respond in a positive way to anything new, especially when it could possibly undermine the stability of the faith. It was also perfectly willing—all too willing—to use its coercive force. In the thirteenth century the Church had also been presented with a major challenge in the revival of Aristotle and pagan philosophy, and it had been able to respond creatively to it as it had in the early Church. But this was not to be so with Copernicanism and the New Science. While there were many clerics who supported Galileo (as there were many laymen who opposed him), the vast majority of the Church's intellectual and institutional leaders were incapable of perceiving the proper boundaries of faith and reason, of the natural and the supernatural, and seemingly indifferent and untouched by the Church's own rich tradition of how to approach these realities intelligently. It created an unfortunate and dangerous breach between the Church and the culture, between the Church and the mind of the age; and while for some that breach was only nominal, a formal and verbal acceptance of the condemnation even while undermining it—a most bizarre thing for an organization whose main stock in trade is the truth that will set you free—the breach was to become a real, permanent, and ever-growing one so

[2] Richard Westfall, *Essays on the Trial of Galileo* (Notre Dame, Ind.: Vatican Observatory Publications, 1989), 33.

that much of modern mankind assumes the utter incompatibility of the two.

Finally, there is Pope Urban VIII, the man to whom, ultimately, the responsibility for the condemnation of Galileo must be placed. More than any pope before him, he controlled the apparatus of the Inquisition, and if the will of the prince is law (as the saying goes), this was particularly true of him in relation to the Inquisition, which was dominated by men owing loyalty to him. His need for control and subservience, which are the perennial temptations of the absolute monarch, did the Church no good and have left a mark against it that has lasted to the present day. While we can understand why he felt betrayed by Galileo and his friends, the intensity and implacability of his response were completely disproportionate to the situation and reveal a personality far too vain and enamored of self to be a truly worthy occupant of the throne of Saint Peter. Even if this extreme response was rooted, as I believe, in his genuinely held skeptical principles, the pope's personal opinions should not guide the running of the Church when so much is at stake. As a cardinal he had worked against a condemnation of Galileo and Copernicus, not because he believed in their truthfulness, but because he believed that their theories could never be proven, were incapable, even, of being proven. He would have been better off if he had taken his own advice as pope.

The skepticism of an Urban VIII was abhorrent to Galileo. It not only makes an investigation of nature impossible, makes science impossible, but it is an insult to God's creation which is a sign from above, a manifestation of God's wisdom and glory. Galileo's rejection of such a worldview was the unmentioned charge against him. Galileo's view is not new, and it has a venerable history both before and after him. Pope Benedict XVI, in a letter to a Galileo congress in 2009, commended him for his insight that the world has an intelligibility that can speak to the human mind and which points out a way that goes beyond mere phenomena to God himself, who is the Supreme Intelligence.

Matter has an intelligibility that can speak to the human mind and point out a way that goes beyond the mere phenomenon. It is Galileo's lesson which led to this thought. Was it not the Pisan scientist who maintained that God wrote the book of nature in the language of mathematics? Yet the human mind invented mathematics in order to understand creation; but if nature is really structured with a mathematical language and mathematics invented by man can manage to understand it, this demonstrates something extraordinary. The objective structure of the universe and the intellectual structure of the human being coincide; the subjective reason and the objectified reason in nature are identical. In the end it is "one" reason that links both and invites us to look to a unique creative Intelligence.[3]

In the introduction I mentioned that the conflict thesis had taken on a new life since the 1990s, and particularly by Catholic writers. This trend has even been noticed by other scholars.[4] Since it had been the general tendency among Catholic scholars to defend vehemently the complementarity and harmony of faith and science, and equally vehemently to downplay the possible conflict, this was quite striking. Thus Richard Blackwell, a professor of philosophy at the Jesuit-run St. Louis University, argued in a 1991 book that the centralizing tendency of the Counter-Reformation Church, the focus on authority and obedience which are the key elements in the logic of centralizing power, was at the heart of the Galileo Affair.

In effect, centrally institutionalized authority tends to evolve into power. Human frailty being what it is, the potential for abuse increases. We begin to see an emphasis on obedience rather than rational evaluation, on tests of faith, on loyalty oaths, on intimidation, on secret proceedings, on unnamed accusers and unspecified allegations, on the use of the courts to suppress recalcitrants—and ultimately on the whole repertoire of the Inquisition. This is not a fantasy scenario. Rather it is precisely what happened in the Galileo affair.[5]

[3] Benedict XVI, Message of His Holiness Benedict XVI to Archbishop Rino Fisichella on the Occasion of the International Congress, "From Galileo's Telescope to Evolutionary Cosmology. Science, Philosophy and Theology in Dialogue", November 26, 2009, http://www.vatican.va/holy_father/benedict_xvi/messages/pont-messages/2009/documents/hf_ben-xvi_mes_20091126_fisichella-telescopio_en.html.

[4] Rivka Feldhay, "Recent Narratives on Galileo and the Church; or The Three Dogmas of the Counter-Reformation", in *Galileo in Context*, ed. Jurgen Renn (Cambridge: Cambridge University Press, 2001), 219–37.

[5] Richard J. Blackwell, *Galileo, Bellarmine, and the Bible* (Notre Dame, Ind.: Notre Dame University Press, 1991), 176–77.

Further, in a 1998 essay, when discussing whether there could be another Galileo case, he again pointed to the centralizing tendency of the Catholic Church (which had only become much more centralized since the definition of Papal Infallibility in 1870) as a major difficulty.[6] While the ideal of harmony could not be argued with, he stated, yet in the practical order, because of their very different methodologies, conflict between science and revelation was quite possible, and the same forces that were in play in Galileo's time are in play today. Religious authority in the Catholic tradition, he believed, is "monolithic, centralized, esoteric, resistant to change, and self-protective", while authority in science is "pluralistic, democratic, public, fallibilistic and self-corrective". "Their methodologies can and do lead to conflict. That is the root of the problem." He believed that this difficulty was further magnified by the fact that the present training of theologians and scientists "seems almost designed to exaggerate and to perpetuate the gulf between the two mindsets".[7] He also had grave doubts whether the Church even now was as committed to intellectual honesty and freedom of thought as it ought to be, and therefore was more open to another Galileo case.

While the logic of centralized authority within the Catholic Church, with its lack of transparency and accountability, checks and balances, is certainly a very serious danger and has made for the possibility of another Galileo case, any organization, particularly one that makes dogmatic claims, has many of the same dangers. There also seems to be among some Catholic authors even a sense of inevitability about conflict and that both in Galileo's condemnation and in possible newer cases the "authoritarian" nature of the Church is the main culprit. It is to be wondered if this new awareness is not, perhaps, the consequence of the greater stress on doctrinal orthodoxy and regular discipline in the reigns of recent popes than in some profound insight into the nature and the history of the Church herself. They also seem somewhat too sanguine and idealistic about science itself. Certainly, the history of science is not without its share of Grand Inquisitors and show trials, of rigidity, dogmatism, and intolerance of other opinions.

---

[6] Richard J. Blackwell, "Could There Be Another Galileo Case?", in *The Cambridge Companion to Galileo*, ed. Peter Machamer (Cambridge: Cambridge University, 1998), 348–66.

[7] Ibid., 351, 358–59.

Nor are the baser human vices unknown to it, including the falsification of research, empire building, and self-aggrandizement.

It might rather be more useful to look at a more secular figure, Marcello Pera, who has also promoted a more conflictual view of things.[8] Pera, a respected Italian philosopher of science and president of the Italian Senate from 2001 to 2006, has even written a book with Cardinal Ratzinger (later Pope Benedict XVI).[9] He more recently defended the Judeo-Christian roots of free societies though he himself is an atheist.[10] He believes that "the fire of new Galileo affairs is still smoldering under the ashes that were thought to be cold" and that such cases "do not depend on historical circumstances, the imprudence of men, the transition from one tradition to another, or the power and prerogatives of institutions".[11] They are in fact constitutive. While he acknowledges that the acceptance of the "independence principle" by the Church (that there are two different realms of knowledge, religious and secular) has, in historical practice, favored both science and religion, protecting them from mutual conflict, he believes that it has not eliminated that possibility since inevitably the two realms overlap. According to the independence principle, conflict can never happen because science deals with questions of facts and theology with questions of salvation. The problem is, he states, "Are there factual questions that a Catholic believer takes (or has to take) as essential for his or her salvation?" The answer is clearly yes. There are dogmas such as the virginity of Mary or Christ's bodily resurrection, which are beliefs about physical facts; and there are also many other beliefs essential to the faith or closely associated with it. Pera correctly points out that the Church cannot give up on certain facts. An incarnational religion is going to be hopelessly enmeshed with all sorts of physical facts, and since we are living in the world, our choices will touch upon real physical things. While a great many

---

[8] Marcello Pera, "The God of Theologians and the God of Astronomers: An Apology of Bellarmine", in *The Cambridge Companion to Galileo*, ed. Peter Machamer (Cambridge: Cambridge University Press, 1998), 367–87.

[9] Joseph Ratzinger and Marcello Pera, *Without Roots: The West, Relativism, Christianity, and Islam* (New York: Basic Books, 2006).

[10] Marcello Pera, *Why We Should Call Ourselves Christians: The Religious Roots of Free Societies* (New York: Encounter Books, 2011).

[11] Pera, "The God of Theologians and the God of Astronomers", 385.

issues are clearly separable from salvation questions, a great many are not. Pera believes that these latter things might have to be changed if science demonstrated something opposed to them, but he also makes the important point that our understanding of modern science itself makes all of the "truths" of science hypothetical. Blackwell has also briefly touched upon this important question.[12]

If scientific laws are fallible and, therefore, scientific teaching can never be proven *conclusively*, what are we to do? For the believer there is the question of what to do with the Augustinian principle of exegesis that a literal reading of Scripture should defer to a proven fact of science. Never to defer would be excessive, but what degree of probability would be needed?

There is an old maxim that goes back to Cicero, though the idea predates him: "Historia magistra vitae" (History is the teacher of life). History really is the great teacher of life, and from the complexity and messiness of the Galileo Affair, I hope one has gained not only a better understanding of what happened but also, as Jacob Burckhardt wrote so many years ago, made oneself not merely "shrewder (for next time), but wiser (for ever)".[13] One thing that one can certainly take from this story is the need for caution and restraint in one's judgments, particularly in reference to the great questions; and one can do no worse, and probably no better, than to follow the advice of Saint Augustine (which Galileo quoted in the *Letter to the Grand Duchess Christina*):

> Meanwhile we should always observe that restraint that is proper to a devout and serious person and on an obscure question entertain no rash belief. Otherwise, if evidence later reveals the explanation we are likely to despise it because of our attachment to our error, even though this explanation may not be in any way opposed to the sacred writings of the Old or New Testament.[14]

The official Church's opinion of Galileo has moved a great deal from the trial of 1633. When we look at the exalted and even exultant

[12] Richard J. Blackwell, *Science, Religion and Authority: Lessons from the Galileo Affair* (Milwaukee, Wis.: Marquette University Press, 1998), 65, n. 32.

[13] Jacob Burckhardt, *Reflections on History* (Indianapolis, Ind.: Library Classics, 1979), 39.

[14] Saint Augustine, *The Literal Meaning of Genesis*, trans. John Hammond Taylor, 2 vols. Ancient Christian Writers (New York: Paulist Press, 1982), 1:73.

language about Galileo from Pope Benedict XVI and other major Vatican figures (one curial official even stated that Galileo could become for some the ideal patron for the dialogue between faith and reason), we see that he has clearly moved from heretic to hero. Looking at the trajectory of Galileo's relationship with the Church, the historian J. L. Heilbron has written that it is only a matter of time before he is canonized as a saint. (*Sancte Galileo Galilei ... Ora pro nobis?*) Already in his 1999 book on cathedrals as solar observatories (*The Sun in the Church*) he prophesized that Galileo would no doubt be shortly—within a hundred years or so—canonized a saint. More recently, in his 2010 biography of Galileo, he returned to the topic more expansively though with a longer time frame. "According to Galileo's mechanics, the slightest force can move the greatest weight given sufficient time. The direction is clear. Who can doubt that within another 400 years the church will recognize Galileo's divine gifts, atone for his sufferings, ignore his arrogance, and make him a saint?" As he points out, there are already relics galore: a right thumb, two fingers, and a tooth are reverently displayed at the Museo Galileo in Florence along with various other personal objects such as one of his telescopes and the original but broken lens from the telescope with which he first saw the Medicean Stars in 1610. To the objection that Galileo performed no miracles, he refers to the stupendous miracle of what he achieved: raising the earth to the heavens, making planets so many earths, revealing that our moon was not unique in the universe. "Then there was the miracle of himself, a rare combination of talents and personalities, who, despite mania and depression, arthritis, gout, hernias, blindness, and overindulgence in wine and wit, lived to write three books–the *Messenger*, the *Dialogue*, and the *Discourse*—any one of which would have given him enduring fame.... What then were the miracles of Thomas Aquinas?"[15]

In fact, while it is true that Pope John XXII, who canonized Thomas Aquinas in 1323, once stated that Aquinas' doctrine alone was a miracle, at Aquinas' tomb a prodigious number of miracles did occur, embarrassingly so for so cerebral a saint, which goes a long way to explain his rapid canonization. While Galileo has been praised, by Pope Benedict XVI for example, as a model of how

[15] John L. Heilbron, *The Sun in the Church* (Cambridge, Mass.: Harvard University Press, 1999), 211, and Heilbron, *Galileo*, 365.

science should be done, of the integral unity of faith and reason, no one has ever claimed him as a model of holiness or heroic virtue—which one should expect of a saint. And while Heilbron may see physical miracles as something inferior and unnecessary when compared to intellectual achievements, still, there is no clearer indication of God's favor and friendship, and therefore of sanctity, than these.

BIBLIOGRAPHY

Abetti, Giorgio. "Galileo the Astronomer". *Popular Astronomy* 159 (1951): 138–43.

Alberigo, Giuseppe. "From Council of Trent to 'Tridentinism'". In *From Trent to Vatican II: Historical and Theological Investigations*, edited by Raymond F. Bulman and Frederick J. Parrella, 19–37. New York: Oxford University Press, 2006.

Artigas, Mariano, William Shea, and Rafael Martinez. "New Light on the Galileo Affair?" In *The Church and Galileo*, edited by Ernan McMullin, 213–33. Notre Dame, Ind.: Notre Dame University Press, 2005.

Augustine. *The Literal Meaning of Genesis*. Translated by John Hammond Taylor. 2 vols. Ancient Christian Writers, 41–42. New York: Paulist Press, 1982.

Baldini, Ugo. "The Academy of Mathematics of the Collegio Romano from 1553 to 1612". In *Jesuit Science and the Republic of Letters*, edited by Mordechai Feingold, 47–98. Cambridge, Mass.: M.I.T. Press, 2003.

————. "The Roman Inquisition's Condemnation of Astrology: Antecedents, Reasons and Consequences". In *Church Censorship and Culture in Early Modern Italy*, edited by Gigliola Fragnito, translated by Adrian Belton, 79–110. Cambridge: Cambridge University Press, 2001.

Barbieri, Edoardo. "Tradition and Change in the Spiritual Literature of the Cinquecento". In *Church Censorship and Culture in Early Modern Italy*, edited by Gigliola Fragnito, translated by Adrian Belton, 111–33. Cambridge: Cambridge University Press, 2001.

Barker, Peter. "Stoic Contributions to Early Modern Science". In *Atoms, Pneuma and Tranquility: Epicurean and Stoic Themes in European Thought*, edited by Margaret J. Osler, 135–54. Cambridge: Cambridge University Press, 1991.

Benedict XVI. "Meeting with the Parish Priests and Clergy of Rome, Feb. 14, 2013". *L'Osservatore Romano*, English edition, February 20, 2013, 8.

———. Message of His Holiness Benedict XVI to Archbishop Rino Fisichella on the Occasion of the International Congress, "From Galileo's Telescope to Evolutionary Cosmology. Science, Philosophy and Theology in Dialogue", November 26, 2009. http://www.vatican.va/holy_father/benedict_xvi/messages /pont-messages/2009/documents/hf_ben-xvi_mes_2009 1126_fisichella-telescopio_en.html.

Beretta, Francesco. "The Documents of Galileo's Trial: Recent Hypotheses and Historical Criticism". In *The Church and Galileo*, edited by Ernan McMullin, 191–212. Notre Dame, Ind.: Notre Dame University Press, 2005.

———. "Galileo, Urban VIII, and the Prosecution of Natural Philosophers". In *The Church and Galileo*, edited by Ernan McMullin, 234–61. Notre Dame, Ind.: Notre Dame University Press, 2005.

Bianchi, Luca. "Continuity and Change in the Aristotelian Tradition". In *The Cambridge Companion to Renaissance Philosophy*, edited by James Hankins, 49–71. Cambridge: Cambridge University Press, 2007.

Black, Christopher F. *The Italian Inquisition*. New Haven, Conn.: Yale University Press, 2009.

Blackwell, Richard J. *Behind the Scenes at Galileo's Trial*. Notre Dame, Ind.: Notre Dame University Press, 2006.

———. "Could There Be Another Galileo Case?" In *The Cambridge Companion to Galileo*, edited by Peter Machamer, 348–66. Cambridge: Cambridge University Press, 1998.

———. *Galileo, Bellarmine, and the Bible*. Notre Dame, Ind.: Notre Dame University Press, 1991.

———. Introduction to *A Defence of Galileo, the Mathematician from Florence*, by Thomas Campanella, 1–34. Translated by Richard J. Blackwell. Notre Dame, Ind.: University of Notre Dame Press, 1994.

———. *Science, Religion and Authority: Lessons from the Galileo Affair*. Milwaukee, Wis.: Marquette University Press, 1998.

Blair, Anne, and Anthony Grafton. "Reassessing Humanism and Science". *Journal of the History of Ideas* 53 (1992): 535–40.

Borromeo, Agostino. "The Inquisition and Inquisitorial Censorship". In *Catholicism in Early Modern History: A Guide to Research*, edited by John O'Malley, 253–72. St. Louis: Center for Reformation Studies, 1988.

Bouwsma, William J. *Venice and the Defense of Republican Liberty*. Berkeley: University of California Press, 1968.

Broderick, James. *The Life and Work of Blessed Robert Francis Cardinal Bellarmine, S.J., 1542–1621*. 2 vols. New York: P.J. Kenedy & Sons, 1928.

Brooke, John, and Geoffrey Cantor, *Reconstructing Nature: The Engagement of Science and Religion*. Edinburgh: T&T Clark, 1998.

Brooke, John, and Ian Maclean, eds. *Heterodoxy in Early Modern Science and Religion*. Oxford: Oxford University Press, 2005.

Brotóns, Victor Navarro. "The Reception of Copernicus in Sixteenth-Century Spain: The Case of Diego de Zúñiga". *Isis* 86 (1995): 52–78.

Brucker, Gene A. "The Medici in the Fourteenth Century". *Speculum* 32, no. 1 (January 1957): 1–26.

Bucciantini, Massimo, and Michele Camerota. "Once More About Galileo and Astrology: A Neglected Testimony". *Galilaeana* II (2005): 229–32.

Burckhardt, Jacob. *The Civilization of the Renaissance in Italy*. Translated by S. G. C. Middlemore. New York: Random House, 1954.

―――. *Reflections on History*. Indianapolis, Ind.: Liberty Classics, 1979.

Burke, Peter. "Early Modern Venice as a Center of Information and Communication". In *Venice Reconsidered: The History and Civilization of an Italian City-State, 1297–1797*, edited by John Martin and Dennis Romano, 389–419. Baltimore: Johns Hopkins University Press, 2000.

Byrne, James Steven. "A Humanist History of Mathematics? Regiomontanus's Padua Oration in Context". *Journal of the History of Ideas* 67, no. 1 (January 2006): 41–61.

Campanella, Thomas. *A Defence of Galileo the Mathematician from Florence*. Translated by Richard J. Blackwell. Notre Dame, Ind.: University of Notre Dame Press, 1994.

Capponi, Niccolo. "Le Palle di Marte: Military Strategy and Diplomacy in the Grand Duchy of Tuscany under Ferdinand II de'

Medici (1621–1670)". *The Journal of Military History* 68 (2004): 1105–41.

Celenza, Christopher. "The Revival of Platonic Philosophy". In *The Cambridge Companion to Renaissance Philosophy*, edited by James Hankins, 72–96. Cambridge: Cambridge University Press, 2007.

Chadwick, Henry. *Early Christian Thought and the Classical Tradition: Studies in Justin, Clement and Origen*. Oxford: Clarendon Press, 1966.

Chambers, David, and Brian Pullan, eds. *Venice: A Documentary History, 1450–1630*. Blackwells: London, 1992.

Chojnacki, Stanley. "Identity and Ideology in Renaissance Venice: The Third *Serrata*". In *Venice Reconsidered: The History and Civilization of an Italian City-State 1297–1797*, edited by John Jeffries Martin and Dennis Romano, 263–94. Baltimore: Johns Hopkins University Press, 2002.

Cipolla, Carlo M. *Guns and Sails in the Early Phase of the European Expansion, 1400–1700*. London: Collins, 1965.

Cochrane, Eric. "Counter Reformation or Tridentine Reformation? Italy in the Age of Carlo Borromeo". In *San Carlo Borromeo: Catholic Reform and Ecclesiastical Politics in the Second Half of the Sixteenth Century*, edited by John M. Headley and John B. Tomaro, 31–46. Washington, D.C.: Folger Books, The Folger Shakespeare Library, 1988.

―――. *Florence: The Forgotten Centuries 1527–1800*. Chicago: University of Chicago Press, 1973.

―――, ed. *The Late Italian Renaissance 1525–1630*. London: Macmillan, 1970.

―――. "Science and Humanism in the Italian Renaissance". *The American Historical Review* 81 (1976): 1039–57.

Cole, Janie. "Cultural Clientelism and Brokerage Networks in Early Modern Florence and Rome: New Correspondence between the Barberini and Michelangelo Buonarroti the Younger". *Renaissance Quarterly* 60 (2007): 729–88.

Consolmagno, Guy. "Precursors of Evil". In *Brother Astronomer: Adventures of a Vatican Scientist*, 61–79. New York: McGraw-Hill, 2000.

Constant, Eric A. "A Reinterpretation of the Fifth Lateran Council Decree *Apostoli regiminis* (1513)". *The Sixteenth Century Journal* 33 (2002): 353–76.

Copenhaver, Brian P. "How to Do Magic, and Why: Philosophical Prescriptions". In *The Cambridge Companion to Renaissance Philosophy*, edited by James Hankins, 137–69. Cambridge: Cambridge University Press, 2007.

Copernicus, Nicholas. *On The Revolutions*. Edited by Jerzy Dobrzycki. Translation and commentary by Edward Rosen. Baltimore: Johns Hopkins University Press, 1978.

Coryat, Thomas. *Coryat's Crudities*. 2 vols. Glasgow: James MacLehose and Sons, 1900.

Coyne, George V. "The Church's Most Recent Attempt to Dispel the Galileo Myth". In *The Church and Galileo*, edited by Ernan McMullin, 340–59.

Coyne, G. V., and U. Baldini. "The Young Bellarmine's Thoughts on World Systems". In *The Galileo Affair: A Meeting of Faith and Science*, edited by G. V. Coyne, M. Heller, and J. Zycinski, 103–10. Vatican City, 1985; and in *The Louvain Lectures of Bellarmine and the Autograph Copy of His 1616 Declaration to Galileo*. Vatican City: Vatican Observatory, 1984.

Crawford, Francis Marion. *Salve Venetia: Gleanings from Venetian History*. 2 vols. New York: Macmillan, 1905.

Crombie, Alistair. "Sources of Galileo's Early Natural Philosophy". In *Science, Art and Nature in Medieval and Modern Thought*, 149–63. London: Hambledon Press, 1996.

Dandelet, Thomas James. "Politics and State System after the Habsburg-Valois Wars". In *Early Modern Italy: 1550–1796*, edited by John Marino, 11–29. Oxford: Oxford University Press, 2002.

———. *Spanish Rome 1500–1700*. New Haven, Conn.: Yale University Press, 2001.

D'Argaville, Brian T. "Inquisition and Metamorphosis: Paolo Veronese and the 'Ultima Cena' of 1573". *RACAR: revue d'art canadienne/Canadian Art Review* 16, no. 1 (1989): 43–48, 99.

Davis, Edward. "Christianity and Early Modern Science: The Foster Thesis Reconsidered". In *Evangelicals and Science in Historical Perspective*, edited by David N. Livingstone, D. G. Hart, and Mark A. Noll, 75–95. Oxford: Oxford University Press, 1999.

Davis, James Cushman. *The Decline of the Venetian Nobility as a Ruling Class*. Baltimore: Johns Hopkins University Press, 1962.

Dear, Peter. *Christian Slaves, Moslem Masters: White Slavery in the Mediterranean, the Barbary Coast, and Italy, 1500–1800.* New York: Palgrave Macmillan, 2004.

———. *Mersenne and the Learning of the Schools.* Ithaca, N.Y.: Cornell University Press, 1988.

De Ceglia, Francesco Paolo. "Additio Illa Non Videntur Edenda: Giuseppe Biancani, Reader of Galileo in an Unedited Censored Text". In *The New Science and Jesuit Science: Seventeenth-Century Perspectives,* edited by Mordechai Feingold, 159–86. Dordrecht: Kluwer, 2003.

D'Elia, Pasquale. "The Spread of Galileo's Discoveries in the Far East (1610–1640)", *East and West* 1, no. 3 (October 1950): 56–63.

Delumeau, Jean. "Rome: Political and Administrative Centralization in the Papal State in the Sixteenth Century". In *The Late Italian Renaissance: 1525–1630,* edited by Eric Cochrane, 287–304. London: Macmillan, 1970.

De Roover, Raymond. *The Rise and Decline of the Medici Bank, 1397–1494.* Cambridge, Mass.: Harvard University Press, 1963.

Desmond, Adrian. *Huxley: From Devil's Disciple to Evolution's High Priest.* Reading, Mass.: Addison-Wesley, 1997.

Donnelly, John. "The Jesuit College at Padua: Growth, Suppression, Attempts at Restoration". *Archivum Historicum Societatis Iesu* 50 (1982): 45–79.

Dooley, Brendan. *Morandi's Last Prophecy and the End of Renaissance Politics.* Princeton, N.J.: Princeton University Press, 2002.

Drake, Stillman. *Discoveries and Opinions of Galileo.* New York: Anchor Books, 1957.

Drake, Stillman, and C.D. O'Malley. *The Controversy on the Comets of 1618.* Philadelphia: University of Pennsylvania Press, 1960.

Fantoli, Annibale. "Astrology, Religion, and Politics in Counter-Reformation Rome". In *Science, Culture, and Popular Belief in Renaissance Europe,* edited by Stephen Pumfrey, Paolo L. Rossi, and Maurice Slawinski, 249–73. Manchester: Manchester University Press, 1991.

———. *Galileo for Copernicus and for the Church.* Translated by G.V. Coyne. 2nd. ed. Vatican City: Vatican Observatory, 1996.

Feingold, Mordechai. "The Grounds for Conflict: Grienberger, Grassi, Galileo, and Posterity". In *The New Science and Jesuit*

Science: Seventeenth-Century Perspectives, edited by Mordechai Feingold, 121–57. Dordrecht: Kluwer, 2003.

————, ed. *Jesuit Science and the Republic of Letters*. Cambridge, Mass.: M.I.T. Press, 2003.

————. "Jesuits: Savants". In *Jesuit Science and the Republic of Letters*, edited by Mordechai Feingold, 1–45. Cambridge, Mass.: M.I.T. Press, 2003.

————, ed. *The New Science and Jesuit Science: Seventeenth-Century Perspectives*. Dordrecht: Kluwer, 2003.

Feldhay, Rivka. "The Cultural Field of Jesuit Science". In *The Jesuits: Cultures, Science, and the Arts 1540–1773*, edited by Gauvin Alexander Bailey, Steven J. Harris, John W. O'Malley, and T. Frank Kennedy, 105–30. Toronto: University of Toronto, 1999.

————. *Galileo and the Church*. Cambridge: Cambridge University Press, 1995.

————. "Recent Narratives on Galileo and the Church; or The Three Dogmas of the Counter-Reformation". In *Galileo in Context*, edited by Jurgen Renn, 219–37. Cambridge: Cambridge University Press, 2001.

Ferrone, Vincenzo, and Massimo Firpo. "From Inquisitors to Microhistorians: A Critique of Pietro Redondi *Galileo eretico*". *Journal of Modern History* 58 (1986): 485–524.

Ficino, Marsilio. *Three Books on Life*. Translation with introduction and notes by Carol V. Kaske and John R. Clark. Binghamton, N.Y.: Medieval and Renaissance Texts and Studies, 1989.

Finocchiaro, Maurice A. *The Galileo Affair*. Berkeley: University of California Press, 1989.

————. *Retrying Galileo, 1633–1992*. Berkeley: University of California Press, 2005.

Fragnito, Gigliola. "Cardinals' Courts in Sixteenth Century Rome". *Journal of Modern History* 65 (1993): 26–56.

————. "The Central and Peripheral Organization of Censorship". In *Church Censorship and Culture in Early Modern Italy*, edited by Gigliola Fragnito, translated by Adrian Belton, 13–49. Cambridge: Cambridge University Press, 2001.

————, ed. *Church Censorship and Culture in Early Modern Italy*. Translated by Adrian Belton Cambridge: Cambridge University Press, 2001.

————. "The Expurgatory Policy of the Church and the Works of Gasparo Contarini". In *Heresy, Culture, and Religion in Early Modern Italy: Contexts and Contestations,* edited by Ronald K. Delph, Michelle M. Fontaine, and John Jeffries Martin, 193–210. Kirksville, Mo.: Truman State University Press, 2006.

Funkenstein, Amos. *Theology and the Scientific Imagination: From the Middle Ages to the Seventeenth Century.* Princeton, N.J.: Princeton University Press, 1986.

Furnell, F. "Jacopo Mazzini and Galileo". *Physis* 14 (1972): 273–94.

Galilei, Galileo. *Dialogue on the Two Chief World Systems.* Translated by Stillman Drake. New York: Modern Library, 2001.

————. *On the World Systems: New Abridged Translation and Guide.* Translated and edited by Maurice Finocchiaro. Berkeley: University of California Press, 1997.

Garin, Eugenio. *Astrology in the Renaissance: The Zodiac of Life.* Translated by Carolyn Jackson and June Allen. London: Routledge & Kegan Paul, 1983.

Gassendi, Pierre. *Concerning Happiness.* Translated by Erik Anderson. Accessed August 4, 2014. http://www.epicurus.info/etexts /gassendi_concerninghappiness.html#1.3.

Gilley, Sheridan, and Ann Loades. "Thomas Henry Huxley: The War between Science and Religion". *Journal of Religion* 61 (1981): 285–308.

Gingerich, Owen. *The Book That Nobody Read: Chasing the Revolutions of Nicolaus Copernicus.* New York: Walker & Company, 2004.

————. "The Censorship of Copernicus' *De revolutionibus*". *Journal of the American Scientific Affiliation* (March 1981): 58–60.

————. "Johannes Kepler". *Dictionary of Scientific Biography.* Edited by Charles Coulston Gillispie, 7:289–312. New York: Charles Scribner's Sons, 1973.

————. "Show Trial?" *The American Scholar* 59, no. 2 (Spring 1990): 310–14.

Gleason, Elisabeth. "Who Was the First Counter-Reformation Pope?". *Catholic Historical Review* 81 (1995): 173–84.

Goethe, Johann Wolfgang von. *Goethe's Travels in Italy.* London: George Bell and Sons, 1885.

Goodman, David. *Spanish Naval Power 1589–1665: Reconstruction and Defeat.* Cambridge: Cambridge University Press, 2003.

Gorman, Michael John. "Mathematics and Modesty in the Society of Jesus: The Problems of Christoph Grienberger". In *The New Science and Jesuit Science: Seventeenth-Century Perspectives,* edited by Mordechai Feingold, 1–120. Dordrecht: Kluwer, 2003.

Grafton, Anthony. *Cardano's Cosmos: The Worlds and Works of a Renaissance Astrologer.* Cambridge, Mass.: Harvard University Press, 1999.

Graham-Dixon, Andrew. *Caravaggio: A Life Sacred and Profane.* New York: W. Norton & Company, 2010.

Grant, Edward. *The Foundations of Modern Science in the Middle Ages: Their Religious, Institutional and Intellectual Contexts.* Cambridge: Cambridge University Press, 1996.

————. *Science and Religion, 400 B.C. to A.D.: From Aristotle to Copernicus.* Westport, Conn.: Greenwood Press, 2004.

————"Science and Theology in the Middle Ages". In *God and Nature: Historical Essays on the Encounter between Science and Religion,* edited by David Lindberg and Ronald Numbers, 49–75. Berkeley: University of California Press, 1986.

————. *A Source Book of Medieval Science.* Cambridge Mass.: Harvard University Press, 1974.

Grendler, Marcella. "Book Collecting in Counter-Reformation Italy: The Library of Gian Vincenzo Pinelli (1535–1601)". *Journal of Library History* 16 (1981): 143–51.

————. "A Greek Collection at Padua: The Library of Gian Vincenzo Pinelli (1535–1601)". *Renaissance Quarterly* 33 (1980): 386–416.

Grendler, Paul. *The Roman Inquisition and the Venetian Press, 1540–1605.* Princeton, N.J.: Princeton University Press, 1977.

————. *The Universities of the Italian Renaissance.* Baltimore: John Hopkins University Press, 2002.

————. "The Universities of the Renaissance and Reformation". *Renaissance Quarterly* 57 (2004): 1–42.

Grubb, James. "Elite Citizens". In *Venice Reconsidered: The History and Civilization of an Italian City-State 1297–1797,* edited by John Jeffries Martin and Dennis Romano, 339–64. Baltimore: Johns Hopkins University Press, 2002.

————. "When Myths Lose Power: Four Decades of Venetian Historiography". *Journal of Modern History* 58 (1986), 43–96.

Guarini, Elena Fasano. "'Rome, Workshop of All the Practices
of the World': From the Letters of Cardinal Ferdinando de'
Medici to Cosimo I and Francesco I". In *Court and Politics in
Papal Rome 1492–1700*, edited by Gianvittorio Signoretto and
Maria Antonietta Visceglia. Cambridge: Cambridge University
Press, 2002.

Guicciardini, Francesco. *The History of Italy*. Translated by Sidney
Alexander. New York: Macmillan, 1969.

Hall, Crystal. *Galileo's Reading*. Cambridge: Cambridge University
Press, 2014.

Hallman, Barbara McClung. *Italian Cardinals, Reform, and the Church
as Property, 1492–1563*. Berkeley: University of California Press,
1985.

Hankins, James, ed. *The Cambridge Companion to Renaissance Philosophy*. Cambridge: Cambridge University Press, 2007.

————. "Galileo, Ficino and Renaissance Platonism". In *Humanism
and Early Modern Philosophy*, edited by Jill Kraye and M. W. F.
Stone, 209–24. London: Routledge, 2000.

————. "The Popes and Humanism". In *Rome Reborn: The Vatican Library and Renaissance Culture*, edited by Anthony Grafton,
47–85. New Haven, Conn.: Yale University Press, 1993.

Headley, John M. *Tommaso Campanella and the Transformation of the
World*. Princeton, N.J.: Princeton University Press, 1997.

Heilbron, John L. "Censorship of Astronomy in Italy after Galileo".
In *The Church and Galileo*, edited by Ernan McMullin, 279–322.
Notre Dame, Ind.: Notre Dame University Press, 2005.

————. *Galileo*. Oxford: Oxford University Press, 2010.

————. *The Sun in the Church*. Cambridge, Mass.: Harvard University Press, 1999.

Hellyer, Marcus. *Catholic Physics: Jesuit Natural Philosophy in Early
Modern Germany*. Notre Dame, Ind.: University of Notre Dame
Press, 2005.

Hendrix, Scott. *How Albert the Great's Speculum Astronomiae Was
Interpreted and Used by Four Centuries of Readers: A Study in Late
Medieval Medicine, Astronomy, and Astrology*. New York: Edwin
Mellen Press, 2010.

Hess, Andrew C. "The Battle of Lepanto and Its Place in Mediterranean History". *Past & Present* 57 (1972): 53–73.

Hollingsworth, Mary, and Carol Richardson, eds. *The Possessions of a Cardinal: Piety, Politics, and Art, 1450–1700*. University Park: Penn State University Press, 2010.

Hsia, R. Po-chia. *The World of Catholic Renewal 1540–1770*. Cambridge: Cambridge University Press, 1998.

Hudon, William H. "Religion and Society in Early Modern Italy— Old Questions, New Insights". *American Historical Review* 101, no. 3 (June 1996): 783–804.

Huemer, Frances. "Rubens's Portrait of Galileo in the Cologne Group Portrait". *Notes in the History of Art* 24, no. 1 (Fall 2004): 18–25.

Ippolito, Antonio Menniti. "The Secretariat of State as the Pope's Special Ministry". In *Court and Politics in Papal Rome, 1492–1700*, edited by Gianvittorio Signorettoand Maria Antonietta Visceglia, 132–57. Cambridge: Cambridge University Press, 2002.

Jaeger, Werner. *Early Christianity and Greek Paideia*. Cambridge, Mass.: Belknap Press of Harvard University Press, 1961.

Kelter, Irving A. "The Refusal to Accommodate: Jesuit Exegetes and the Copernican System". In *The Church and Galileo*, edited by Ernan McMullin, 38–53. Notre Dame, Ind.: Notre Dame University Press, 2005.

Kennedy, Leonard. "Cesare Cremonini and the Immortality of the Human Soul". *Vivarium* 18 (1980): 143–58.

Knox, Bernard. *Oedipus at Thebes*. New Haven, Conn.: Yale University Press, 1957.

Kollerstrom, Nick. "Galileo's Astrology". *The Mountain Astrologer*. Accessed January 24, 2017. http://www.skyscript.co.uk/galast .html#7.

Koyré, Alexandre. "Galileo and the Scientific Revolution of the Seventeenth Century". *The Philosophical Review* 52 (1943): 333–48.

Kraye, Jill. "Philologists and Philosophers". In *The Cambridge Companion to Humanism*, edited by Jill Kraye, 142–60. Cambridge: Cambridge University Press, 1996.

———. "Pietro Pomponazzi (1462–1525) Secular Aristotelianism in the Renaissance". In *Philosophers of the Renaissance*, edited by

Paul Richard Blum, 92–115. Washington, D.C.: Catholic University of America Press, 2010.

———. "The Revival of Hellenistic Philosophies". In *The Cambridge Companion to Renaissance Philosophy*, edited by James Hankins, 97–112. Cambridge: Cambridge University Press, 2007.

Kristeller, Paul Oskar. *Eight Philosophers of the Italian Renaissance*. London: Chatto & Windus, 1965.

———."Florentine Platonism and Its Relation to Humanism and Scholasticism". *Church History* 8 (1939): 201–11.

———. *Renaissance Thought and Its Sources*. Edited by Michael Mooney. New York: Columbia University Press, 1979.

Laird, W.R. "Archimedes among the Humanists". *Isis* 82 (1991): 628–38.

Langdon, Helen. *Caravaggio: A Life*. New York: Farrar, Straus and Giroux, 1998.

Langford, Jerome J. *Galileo, Science and the Church*. Ann Arbor: University of Michigan Press, 1992.

Lattis, James M. *Between Copernicus and Galileo: Christoph Clavius and the Collapse of Ptolemaic Cosmology*. Chicago: University of Chicago Press, 1994.

Lerner, Michel-Pierre. "The Heliocentric 'Heresy' From Suspicion to Condemnation". In *The Church and Galileo*, edited by Ernan McMullin, 11–37. Notre Dame, Ind.: Notre Dame University Press, 2005.

Lindberg, David C. *The Beginnings of Western Science: The European Scientific Tradition in Philosophical, Religious, and Institutional Context, 600 B.C. to A.D. 1450*. Chicago: University of Chicago Press, 1992.

———. "Science and the Early Church". In *God & Nature: Historical Essays on the Encounter between Christianity and Science*, 19–48. Berkeley: University of California Press, 1986.

———. "Galileo, the Church, and the Cosmos". In *When Science & Christianity Meet*, edited by David C. Lindberg and Ronald L. Numbers, 33–60. Chicago: University of Chicago Press, 2003.

Lindberg, David C., and Ronald L. Numbers. "Beyond War and Peace: A Reappraisal of the Encounter between Christianity and Science". *Church History* 55 (1986): 338–54.

Lindberg, David C., and R. S. Westman, eds. *Reappraisals of the Scientific Revolution*. Cambridge: Cambridge University Press, 1990.

Lines, Davis L. "Natural Philosophy in Renaissance Italy: The University of Bologna and the Beginnings of Specialization". *Early Science and Medicine* 6, no. 4 (2001): 267–323.

Lipsius, Justus. *His Second Book of Constancy*. Latin 1584; Englished by John Stradling, 1594; very slightly retouched and annotated by Jan Garrett, 1999–2000. Chap. 6. Accessed January 23, 2017. http://people.wku.edu/jan.garrett/lipsius2.htm.

Lüthy, Christoph, and Leen Spruit. "The Doctrine, Life, and Roman Trial of the Frisian Philosopher Henricus de Veno (1574?–1613)". *Renaissance Quarterly* 56 (2003): 1112–51.

Machamer, Peter, ed. *The Cambridge Companion to Galileo*. Cambridge: Cambridge University Press, 1998.

Marino, John A., ed. *Early Modern Italy 1550–1796*. Oxford: Oxford University Press, 2002.

Martin, John Jeffries. "*Renovatio* and Reform in Early Modern Italy". In *Heresy, Culture, and Religion in Early Modern Italy: Contexts and Contestations*, edited by Ronald K. Delph, Michelle M. Fontaine, and John Jeffries Martin, 1–17. Kirksville, Mo.: Truman State University Press, 2006.

———. *Venice's Hidden Enemies: Italian Heretics in a Renaissance City*. Baltimore: Johns Hopkins University Press, 1993.

Martin, John Jeffries, and Dennis Romano, eds. *Venice Reconsidered: The History and Civilization of an Italian City-State 1297–1797*. Baltimore: Johns Hopkins University Press, 2002.

Mayer, Thomas F. *The Roman Inquisition: A Papal Bureaucracy and Its Laws in the Age of Galileo*. Philadelphia: University of Pennsylvania Press, 2013.

———. "The Roman Inquisition's Precept to Galileo (1616)". *British Journal for the History of Science* 43 (2010): 327–51.

———, ed. *The Trial of Galileo 1612–1633*. North York, Ontario: University of Toronto Press, 2012.

McMullin, Ernan. "The Church's Ban on Copernicanism". In *The Church and Galileo*, edited by Ernan McMullin, 150–90. Notre Dame, Ind.: Notre Dame University Press, 2005.

———, ed. *The Church and Galileo*. Notre Dame, Ind.: Notre Dame University Press, 2005.

————, ed. *Galileo, Man of Science*. New York: Basic Books, 1967.

————. "Galileo on Science and Scripture". In *The Cambridge Companion to Galileo*, edited by Peter Machamer, 271–347. Cambridge: Cambridge University, 1998.

————. "Galileo's Theological Venture". In *The Church and Galileo*, edited by Ernan McMullin, 88–116. Notre Dame, Ind.: Notre Dame University Press, 2005.

Mersenne, Marin. *The Truth of the Sciences: Selections from Book I.* Translated by M. S. Mahoney. Accessed January 23, 2017. https://www.princeton.edu/~hos/mike/texts/mersenne/mersenne.htm.

Montaigne, Michel. *The Complete Essays of Montaigne*. Translated by Donald M. Frame. Stanford, Calif.: Stanford University Press, 1948.

————. *Travel Journal*. Translated by Donald M. Frame. San Francisco: North Point Press, 1983.

Mormando, Franco. *Bernini: His Life and His Rome*. Chicago: University of Chicago Press, 2011.

Moryson, Fynes. *An Itinerary*. 4 vols. Glasgow: James MacLehose and Sons, 1907.

Moss, Jean Dietz. "Galilieo's Letter to the Grand Duchess Christina: Some Rhetorical Considerations". *Renaissance Quarterly* 36, no. 4 (1983): 547–76.

————. *Novelties in the Heavens: Rhetoric and Science in the Copernican Controversy*. Chicago: Chicago University Press, 1993.

Muir, Edward. *The Culture Wars of the Late Renaissance*. Cambridge, Mass.: Harvard University Press, 2007.

Murray, Gilbert. *Five Stages of Greek Religion*. Boston: Beacon Press, 1951.

————. *A History of Ancient Greek Literature*. New York: D. Appleton, 1916.

Newman, William R., and Anthony Grafton. "Introduction: The Problematic Status of Astrology and Alchemy in Premodern Europe". In *Secrets of Nature: Astronomy and Alchemy in Early Modern Europe*, 1–37. Cambridge, Mass.: Harvard University Press, 2002.

————, eds. *Secrets of Nature: Astronomy and Alchemy in Early Modern Europe*. Cambridge, Mass.: Harvard University Press, 2002.

Oakley, Francis. "The Absolute and Ordained Power of God in Sixteenth and Seventeenth Century Theology". *Journal of the History of Ideas* 59 (July 1998): 437–61.

———. *Omnipotence, Covenant, and Order*. Ithaca, N.Y.: Cornell University Press, 1984.

———. *Omnipotence and Promise: The Legacy of the Scholastic Distinction of Powers*. Toronto: Pontifical Institute of Medieval Studies, 2002.

O'Connor, J.J., and E.F. Robertson. "Giuseppe Biancani". School of Mathematics and Statistics at the University of St. Andrews, Scotland. July 2012. http://www-history.mcs.st-andrews.ac.uk /Biographies/Biancani.html.

Olin, John. *The Catholic Reformation: Savonarola to Ignatius*. New York: Loyola Fordham University Press, 1992.

Olschki, Leonardo. "Galileo's Literary Formation". In *Galileo: Man of Science*, edited by Ernan McMullin, 140–59. New York: Basic Books, 1967.

O'Malley, John W., ed. *Catholicism in Early Modern History: A Guide to Research*. St. Louis: Center for Reformation Studies, 1988.

———. *The First Jesuits*. Cambridge, Mass.: Harvard University, 1995.

———. *Trent: What Happened at the Council*. Cambridge, Mass.: Harvard University Press, 2013.

Panofsky, Erwin. "Galileo as a Critic of the Arts: Aesthetic Attitude and Scientific Thought". *Isis* 47, no. 1 (March 1956): 3–15.

Partner, Peter. "Papal Financial Policy in the Renaissance and Counter-Reformation". *Past & Present* 88 (1980): 17–62.

Pastor, Ludwig. *The History of the Popes from the Close of the Middle Ages*. Edited by Frederick Ignatius Antrobus. 2nd ed. St. Louis: B. Herder, 1901.

Patterson, Miles. *Pius IV and the Fall of the Carafa: Nepotism and Papal Authority in Counter-Reformation Rome*. Oxford: Oxford University Press, 2013.

Patterson, Paul A. *Visions of Christ: The Anthropomorphite Controversy of 399 CE*. Tübingen: Mohr Siebeck, 2012.

Pedersen, Olaf. "Galileo and the Council of Trent: Galileo Affair Revisited". *Journal of the History of Astronomy* 14 (1983): 1–29.

Pera, Marcello. "The God of Theologians and the God of Astrono-
     mers: An Apology of Bellarmine". In *The Cambridge Compan-
     ion to Galileo*, edited by Peter Machamer, 367–87. Cambridge:
     Cambridge University Press, 1998.

————. *Why We Should Call Ourselves Christians: The Religious Roots
     of Free Societies*. New York: Encounter Books, 2011.

Pomata, Gianna. "Family and Gender". In *Early Modern Italy 1550–
     1796*, edited by James Marino, 69–86. Oxford: Oxford Univer-
     sity Press, 2002.

Popkin, Richard H. *The History of Scepticism: From Savonarola to Bayle*.
     New York: Oxford University Press, 2003.

Prodi, Paolo. *The Papal Prince: One Body and Two Souls; The Papal
     Monarchy in Early Modern Europe*. Translated by Susan Haskins.
     Cambridge: Cambridge University Press, 1987.

Quinlan-McGrath, Mary. "The Foundation Horoscope(s) of St.
     Peter's Basilica, Rome: Choosing a Time, Changing the *Storia*".
     Isis 92 (2001): 716–41.

Randall, J. H. *The School of Padua and the Emergence of Modern Science*.
     Padova: Editrice Antenore, 1961.

Ratzinger, Joseph, and Marcello Pera. *Without Roots: The West, Rel-
     ativism, Christianity, and Islam* (New York: Basic Books, 2006).

Reeves, Eileen. *Art and Science in the Age of Galileo*. Princeton, N.J.:
     Princeton University Press, 1997.

Reinhard, Wolfgang. "Papal Power and Family Strategy in the Six-
     teenth and Seventeenth Centuries". In *Princes, Patronage and the
     Nobility: The Court at the Beginning of the Modern Age, c. 1450–
     1650*, edited by Ronald G. Asch and Adolf M. Birke, 329–56.
     Oxford: Oxford University Press, 1991.

Renn, Jürgen, ed. *Galileo in Context*. Cambridge: Cambridge Uni-
     versity Press, 2001.

Rietbergen, Peter. *Power and Religion in Baroque Rome: Barberini Cul-
     tural Politics*. Leiden: Brill, 2006.

Robinson, Adam Patrick. *The Career of Cardinal Giovanni Morone
     (1509–1580): Between Council and Inquisition*. Burlington, Vt.:
     Ashgate, 2012.

Rondet, Henri. *The Grace of Christ: A Brief History of the Theology of
     Grace*. Translated by Tad Guzie. New York: Newman Press,
     1967.

Rösch, Gerhard. "The Serrata of the Great Council and Venetian Society, 1286–1323". In *Venice Reconsidered: The History and Civilization of an Italian City-State 1297–1797*, edited by John Jeffries Martin and Dennis Romano, 67–88. Baltimore: Johns Hopkins University Press, 2002.

Rosen, Edward. "Galileo's Misstatements about Copernicus". *Isis* 49 (1958): 319–30.

———. "Was Copernicus' *Revolutions* Approved by the Pope?" *Journal of the History of Ideas* 36 (1975): 531–42.

Ross, James Bruce, and Mary Martin McLaughlin, eds. *The Portable Renaissance Reader*. New York: Viking Press, 1953.

Rowland, Ingrid. *Giordano Bruno: Philospher/Heretic*. New York: Farrar, Straus and Giroux, 2008.

Rummel, Erika. *The Humanist-Scholastic Debate in the Renaissance and Reformation*. Cambridge, Mass.: Harvard University Press, 1995.

Rutkin, H. Darrel. "Astrology". In *The Cambridge History of Science*. Vol. 3, *Early Modern Science*, edited by David C. Lindberg, Mary Jo Nye, Katharine Park, and Roy Porter (Historiker), 541–61. Cambridge: Cambridge University Press, 2008.

———. "Galileo Astrologer: Astrology and Mathematical Practice in the Late-Sixteenth and Early-Seventeenth Century". *Galilaeana* 2 (2005): 107–43.

Santillana, Giorgio de. *The Crime of Galileo*. Chicago: University of Chicago Press, 1955.

Schmitt, Charles. "Experience and Experiment: A Comparison of Zabarella's View with Galileo's in *De Motu*". *Studies in the Renaissance* 16 (1969): 80–138.

———. "The Faculty of Arts at the Time of Galileo". *Physis* 14 (1972): 243–72.

Scott, John Belton. *Images of Nepotism: The Painted Ceilings of Palazzo Barberini*. Princeton, N.J.: Princeton University Press, 1991.

Segrè, Michael. *In the Wake of Galileo*. New Brunswick, N.J.: Rutgers University Press, 1991.

Sella, Domenico. *Italy in the Seventeenth Century*. London: Addison Wesley Longman, 1997.

Sharratt, Michael. *Galileo: Decisive Innovator*. Cambridge: Cambridge University Press, 1994.

————. "Galileo, Scheiner, and the Interpretation of Sunspots". *Isis* 61 (1970): 488–519.

————. "Guidobaldo del Monte: Galileo's Patron, Mentor and Friend". Edition Open Access, Max Planck Research Library for the History and Development of Knowledge. Accessed January 23, 2017. http://www.edition-open-access.de/proceedings /4/5/index.html.

Shea, Michael, and Mariano Artigas. *Galileo in Rome: The Rise and Fall of a Troublesome Genius.* Oxford: Oxford University Press, 2003.

Signorotto, Gianvittorio, and Maria Antonietta Visceglia, eds. *Court and Politics in Papal Rome 1492–1700.* Cambridge: Cambridge University Press, 2002.

Smith, Logan Pearsall. *The Life and Letters of Sir Henry Wotton.* 2 vols. 1907. Reprint, Oxford: Clarendon Press, 1966.

Smoller, Laura Ackerman. *History, Prophecy, and the Stars: The Christian Astrology of Pierre d'Ailly, 1350–1420.* Princeton, N.J.: Princeton University Press, 1994.

Snyder, Jon R. *Dissimulation and the Culture of Secrecy in Early Modern Europe.* Berkeley: University of California Press, 2009.

Sobel, Dava. *Galileo's Daughter: A Historical Memoir of Science, Faith, and Love.* New York: Walker, 1999.

Speller, Jules. *Galileo's Inquisition Trial Revisited.* Frankfurt: Peter Lang, 2008.

Spini, Giorgio. "The Rationale of Galileo's Religiousness". In *Galileo Reappraised*, edited by Carlo L. Golino, 44–66. Berkeley: University of California Press, 1966.

Storey, Tessa. *Carnal Commerce in Counter-Reformation Rome.* Cambridge: Cambridge University Press, 2008.

Swerdlow, Noel. "Galileo's Discoveries with the Telescope and Their Evidence for Copernican Theory". In *Cambridge Companion to Galileo*, edited by Peter Machamer, 244–70. Cambridge: Cambridge University Press, 1998.

Tedeschi, John. *The Prosecution of Heresy: Collected Studies on the Inquisition in Early Modern Italy.* Binghamton, N.Y.: Medieval and Renaissance Text and Studies, 1991.

————. "The Roman Inquisition and Witchcraft: An Early Seventeenth-Century 'Instruction' on Correct Trial Procedure". *Revue de l'histoire des religions* 200 (1983): 163–88.

Tertullian. *On the Prescription of Heretics 7*. Translated by T. Herbert Bindley. London: SPCK, 1914.

Thomas, William. *The History of Italy (1549)*. Ithaca, N.Y.: Cornell University Press, 1963.

Thorndike, Lynn. "The True Place of Astrology in the History of Science". *Isis* 46 (1955): 273–78.

Turner, Frank M. "The Victorian Conflict between Science and Religion: A Professional Dimension". *Isis* 69 (1978): 356–76.

Tutino, Stefania. *Empire of Souls: Robert Bellarmine and the Christian Commonwealth*. Oxford: Oxford University Press, 2010.

Van Helden, Albert. "Galileo and Scheiner on Sunspots: A Case Study in the Visual Language of Astronomy". *Proceedings of the American Philosophical Society* 140 (1996): 358–96.

———. "Galileo, Telescopic Astronomy, and the Copernican System". In *Planetary Astronomy from the Renaissance to the Rise of Metaphysics: Part A: Tycho Brahe to Newton*, edited by René Taton and Curtis Wilson, 81–105. Cambridge: Cambridge University Press, 1989.

———. "The Invention of the Telescope". *Transactions of the American Philosophical Society* 76, no. 4 (1977): 1–67.

Vanden Broeke, Steven. *The Limits of Influence: Pico, Louvain, and the Crisis of Renaissance Astrology*. Boston: Brill, 2003.

Vasoli, Cesare. "The Contribution of Humanism to the Birth of Modern Science". *Renaissance and Reformation* 3 (1979): 1–15.

Visceglia, Maria Antonietta. "Factions in the Sacred College in the Sixteenth and Seventeenth Centuries". In *Court and Politics in Papal Rome 1492–1700*, edited by Gianvittorio Signoretto and Maria Antonietta Visceglia, 99–131. Cambridge: Cambridge University Press, 2002.

Walker, D.P. *Spiritual and Demonic Magic from Ficino to Campanella*. Notre Dame, Ind.: Notre Dame University Press, 1975.

Wallace, William. "The Certitude of Science in Late Medieval and Renaissance Thought". *History of Philosophy Quarterly* 3, no. 3 (July 1986): 281–91.

———. *Galileo and His Sources: The Heritage of the Collegio Romano in Galileo's Science*. Princeton, N.J.: Princeton University Press, 1984.

———. "Galileo's Jesuit Connections and Their Influence on His Science". In *Jesuit Science and the Republic of Letters*, edited by

Mordechai Feingold, 99–126. Cambridge, Mass.: M.I.T. Press, 2003.

———. "Galileo's Pisan Studies in Science and Philosophy". In *Cambridge Companion to Galileo*, edited by Peter Machamer, 27–52. Cambridge: Cambridge University Press, 1998.

Watts, Edward. "Justinian, Malalas and the End of the Athenian Philosophical Teaching in A.D. 529". *Journal of Roman Studies* 94 (2004): 168–82.

———. "Where to Live the Philosophical Life in the Sixth Century? Damascius, Simplicius and the Return from Persia". *Greek, Roman, and Byzantine Studies* 45 (2005): 285–315.

Westfall, Richard S. *Essays on the Trial of Galileo*. Notre Dame, Ind.: Vatican Observatory Publications, 1989.

———. "Science and Patronage: Galileo and the Telescope". *Isis* 76 (1985): 11–30.

Westman, Robert S. *The Copernican Question: Prognostication, Skepticism, and Celestial Order*. Berkeley: University of California Press, 2011.

———. "The Reception of Galileo's 'Dialogue.' A Partial World Census of Extant Copies". NCCS, 329–71.

White, Lynn, Jr. *Medieval Science and Technology: Collected Essays*. Los Angeles: University of California Press, 1978.

Wilson, Peter H. *The Thirty Years War: Europe's Tragedy*. Cambridge, Mass.: Harvard University Press, 2009.

Wisan, Winifred Lovell. "Galileo and God's Creation". *Isis* 77 (1986): 473–86.

Wootton, David. *Galileo: Watcher of the Skies*. New Haven, Conn.: Yale University Press, 2010.

———. *Paolo Sarpi: Between Renaissance and Enlightenment*. Cambridge: Cambridge University Press, 1983.

Zambelli, Paola. *The Speculum Astronomiae and Its Enigma: Astrology, Theology and Science in Albertus Magnus and His Contemporaries*. Boston: Kluwer Academic Publishers, 1992.

INDEX